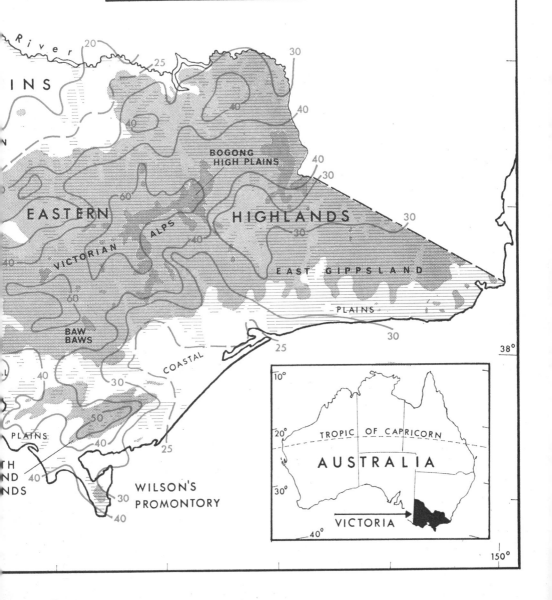

THE MOSQUITOES
OF VICTORIA

THE MOSQUITOES OF VICTORIA

(Diptera, Culicidae)

N. V. DOBROTWORSKY M.Sc., Ph.D.
Senior Research Fellow, Zoology Department,
University of Melbourne

MELBOURNE UNIVERSITY PRESS

LONDON AND NEW YORK : CAMBRIDGE UNIVERSITY PRESS

First published 1965

*Printed and bound in Australia by
Melbourne University Press, Carlton N.3, Victoria*

*Registered in Australia for transmission
by post as a book*

Dewey Decimal Classification Number 595.771
Library of Congress Catalog Card Number 65-21862

Text set in 10 point Times type

ACKNOWLEDGMENTS

I am grateful to Professor M. J. D. White, Zoology Department, University of Melbourne, for his interest and most valuable support when the preparation of this monograph was considered. My thanks are due to those who have assisted me during the preparation of the monograph, particularly to Dr F. H. Drummond, Zoology Department, University of Melbourne, who has undertaken the laborious task of editing the manuscript and making many useful suggestions, and to Dr E. N. Marks, Entomology Department, University of Queensland, for reading and criticizing the manuscript and for many valuable suggestions. I also wish to thank Messrs E. Matthaei and W. McNutt of the Miscroscopy Laboratory, University of Melbourne, for preparing the photographs.

For permission to reproduce illustrations acknowledgments are due to the editors of *Proceedings of The Linnean Society of New South Wales* (a number of figures originally published by the present author and figs. 60, 61 by Dr E. N. Marks), *Journal of the Entomological Society of Queensland* (figs. 46, 47), *Proceedings of the Royal Society of Queensland* (fig. 48) and *University of Queensland Papers* (figs. 39, 58). Details for the end-paper map were taken from the 1944 Report of the Victorian Regional Boundaries Commission.

This book has been published with the assistance of the Scientific Publications Committee of the Commonwealth Treasury.

CONTENTS

Acknowledgments	v
Introduction	3

PART I GENERAL

1 STRUCTURE AND BIOLOGY OF THE ADULT	5
External Anatomy	5
Time of Activity	13
Mating	14
Feeding	15
Oviposition	16
2 STRUCTURE AND BIOLOGY OF THE IMMATURE STAGES	18
Eggs	18
External Anatomy of Larva	20
Biology of Larva	26
Larval Breeding Sites	27
Pupa	28
3 DISTRIBUTION AND COMPOSITION OF THE MOSQUITO FAUNA	30
4 MOSQUITOES AS VECTORS OF DISEASE IN VICTORIA	35

PART II TAXONOMIC

Family Culicidae	39
Subfamily Anophelinae	40
Genus *Anopheles*	41
Subfamily Culicinae	50
Tribe Sabethini	51
Genus *Tripteroides*	51
Tribe Culicini	60
Genus *Mansonia*	60
Genus *Aedeomyia*	67
Genus *Aedes*	70
Genus *Culiseta*	170
Genus *Culex*	191
Appendix	221
References	225
Systematic List of Mosquito Species	231
Index	235

PART I

General

INTRODUCTION

In many parts of the world, mosquitoes are notorious as the vectors of the micro-organisms responsible for malaria, yellow fever, dengue and filariasis. However, these diseases do not occur in Victoria, where, until quite recently, mosquitoes were not known to be of any medical importance. As a consequence they have received far less attention here than in the more northern states of Australia.

The scientific study of Australian mosquitoes began in 1835 when Westwood described *Aedes alternans*. This was followed in 1842 by Erichson's description of *Ae. australis,* and in 1847 Macquart described three more species of the same genus. Serious work on the group in Australia began about 1886 (Marks, 1954). The first important contribution was made by Skuse who in the course of his extensive studies of Australian Diptera, published in the years 1888-90, described thirteen new species of mosquitoes and recorded observations on the biology of many of them. Skuse's material came from New South Wales and Queensland and it was not until 1901 that Theobald described the first species from Victoria. This description, and records of other species in the State, were based on material sent to him by French, the Government Entomologist, and by T. L. Bancroft. T. L. Bancroft, who with his father J. Bancroft pioneered the study of insect-borne diseases in Australia, worked mainly in Queensland but his wide interests in mosquitoes led him to make collecting trips to other states.

The preparation of Theobald's *Monograph of the Culicidae of the World* was initiated as a result of Ross's discovery in 1898 of the role of *Anopheles* in the transmission of malaria. The discovery greatly stimulated research on mosquitoes but the only repercussion in Victoria seems to have been a brief survey, in 1916, of the mosquitoes of the Murray Valley. It was conducted by Taylor on behalf of the Commonwealth Quarantine service, which was concerned lest the large-scale settlement in irrigation areas of ex-servicemen who had contracted malaria while overseas should lead to the establishment of the disease in Victoria. As a result of Taylor's (1917) report that *An. annulipes* was common throughout the area, the Director of Quarantine recommended that 'no man should be allowed to settle in those areas whose record shows that he has suffered whilst on service from malaria or in whose blood the malarial parasites are found'. This recommendation was apparently ignored, an action which, though perhaps unwise at the time, has been vindicated by the complete freedom of the Murray Valley from malaria.

During the years 1924-7 several publications dealing with Australian mosquitoes contained references to Victorian species (Cooling, 1924; Edwards, 1924, 1926; Hill, 1925; Mackerras, 1927a and b). The most important of these was Edwards' (1924) 'Synopsis of the Adult Mosquitoes of the Australasian Region'; it included many new records for Victoria based largely on collecting by Hill. By 1927 a total of twenty-five species had been recorded for the state but four of these are now known to be invalid and the record of *Aedes aegypti* is almost certainly erroneous.

During the next seventeen years little was added to our knowledge of the Victorian mosquito fauna. Research continued in the northern states and was greatly intensified in Queensland during World War II because of the increased danger of malarial outbreaks. A National Mosquito Control Committee was established and entomological units were set up in the Australian Army. A great deal of valuable work was done and two of the resulting publications, *An Atlas of Mosquito Larvae of the Australasian Region* (Lee, 1944) and *The Anopheline Mosquitoes of the Australasian Region* (Lee and Woodhill, 1944), added three new records for Victoria and contained information about the biology of many species occurring in the state.

From this short review, it will be clear that up till the end of the war there had been no systematic study of mosquitoes in Victoria. However, the year 1951 proved to be a turning point. One stimulus was provided by an epidemic of encephalitis in the Murray Valley. This led the Walter and Eliza Hall Institute of Medical Research to organize an intensive study of the mosquito fauna of the Mildura district, during the summer of 1951-2 (Reeves *et al.*, 1954) and in the following years.

A second factor that promoted research on mosquitoes was the introduction of myxoma virus to control wild rabbits. It was soon realized that mosquitoes were the most important vectors of the disease and investigations by officers of the Commonwealth Scientific and Industrial Research Organization and of the Victorian Department of Crown Lands and Survey resulted in extensive collecting of mosquitoes in many parts of the state. As a result of these and other investigations (Dobrotworsky, 1951-62; Marks, 1957-63) the number of species and subspecies recorded from Victoria has risen to a total of seventy-three and, although a number of taxonomic problems remain, it is now possible to give a reasonably complete account of our mosquito fauna.

The first part of the book presents a general account of the structure and biology of mosquitoes. No attempt has been made to review the vast literature on the biology of mosquitoes and most of the references given are to work on species occurring in Victoria. The second part of the book is devoted to a systematic account of the fauna. A description of the adults and fourth-stage larva is given for every species, except for a few in which either the male or larva is unknown, and each description is accompanied by figures illustrating diagnostic details. Figures of wings, legs, the abdomen and mouth appendages are schematic but those of the male terminalia and larvae are accurate camera lucida drawings. In all drawings of the male terminalia, the sternal aspect is shown on the left side, the tergal aspect on the right. The legs are shown in anterior view.

1

STRUCTURE AND BIOLOGY OF THE ADULT

External Anatomy

Mosquitoes belong to the family Culicidae which is subdivided into three subfamilies, two of which occur in Victoria. The subfamily Anophelinae is represented here by a single genus, *Anopheles*; all other genera to which reference is made belong to the subfamily Culicinae.

This account of the external anatomy of mosquitoes will deal mainly with structures used in taxonomic descriptions. Except for venation, where the Comstock-Needham system is adopted, the nomenclature is that used by Edwards (1941). The principal features of an adult mosquito are shown in fig. 1.

HEAD

The lateral parts of the head are occupied by the large compound eyes. The area between and above them, which is called the vertex, is clothed with upright scales which are forked at the broadened apex, and decumbent scales which are either broad and flat, or narrow and curved. These scales often provide useful characters for the identification of species. Anterior to the vertex is the front and beyond this is the clypeus; neither of these parts has much diagnostic value.

The antennae (fig. 2) which arise on either side of the front consist of fifteen segments. The first is small and is concealed by the large, globular second segment, called the torus, which may bear a few scales or hairs, or sometimes both. The remaining segments are narrow and more or less cylindrical; each bears a whorl of hairs, few and short in the female, abundant and long in the male. This sexual dimorphism of the antennae characterizes all Victorian mosquitoes.

The palps (fig. 3) consist of five segments. They also show sexual dimorphism and often provide generic characters. In female Anophelines they are as long as the proboscis but in Culicines only one-eighth to one-sixth as long; in Victoria the only exception is the subgenus *Mucidus* in which the palps are about two-thirds the length of the proboscis. The palps of the female are clothed with scales and short setae. In the males of both subfamilies the palps are usually as long as the proboscis or longer but in *Aedeomyia* and some species of *Tripteroides* they are distinctly shorter. The first three segments comprise the shaft; the penultimate and terminal segments are often swollen and in *Culex* and some *Mansonia* and *Culiseta* they are turned upwards; in other genera they lie parallel to

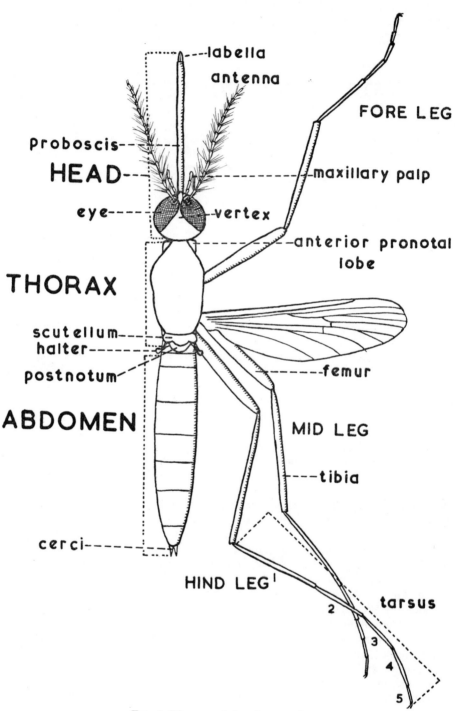

Fig. 1 Diagram of female mosquito.

STRUCTURE AND BIOLOGY OF THE ADULT 7

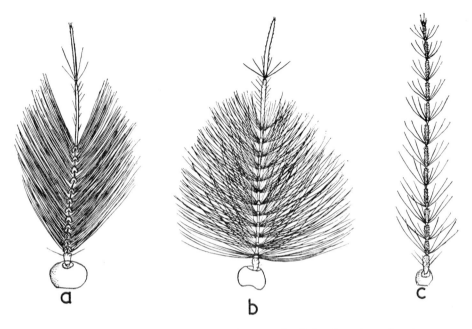

FIG. 2 Antennae of mosquitoes. *Aedes*: a, male. *Culex*: b, male; c, female.

the proboscis. The palps are clothed with scales and usually there are long hairs on the apical part of the shaft and the following segments.

The proboscis consists of the elongated labium and the piercing and feeding structure it ensheathes; at the distal end the labium terminates in a pair of small lobes, the labellae. Scale colour on the proboscis and also on the palps often assists in the identification of species.

THORAX (fig. 4)

The thorax is composed of three segments but, as in all Diptera, the prothorax and metathorax are much reduced by comparison with the mesothorax, the wing-bearing segment. The wall of the thorax is made up of a number of more or less distinct hardened areas, the sclerites, linked by areas of membranous cuticle. The distribution of bristles and scales on the sclerites, membranous areas and appendages (legs, wings and halteres) are of great importance in mosquito taxonomy.

Prothorax. Anteriorly on either side of the prothorax there is a prominent anterior pronotum and, between this and the anterior spiracle, a posterior pronotum with a small spiracular area demarcated by a ridge. The propleuron is a small sclerite just above the base of the anterior coxa and between the two coxae is the prosternum. These sclerites bear bristles and scales; the presence of bristles on the spiracular area distinguishes *Culiseta* and *Tripteroides* from all other genera of mosquitoes found in Victoria.

Mesothorax. The dorsal sclerites of the mesothorax are the scutum, the paired paratergites, the scutellum and the postnotum. The scutum, which is very large, covers most of the dorsal side of the mesothorax; just

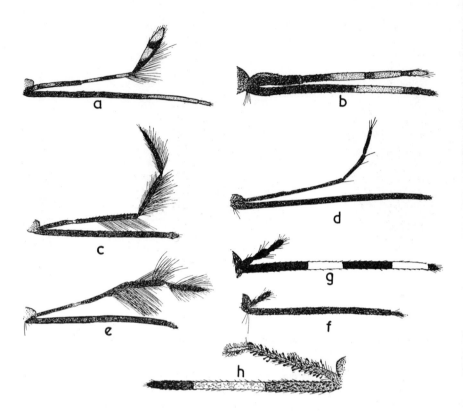

Fig. 3 Proboscis and palp of mosquitoes. *Anopheles annulipes*: a, male; b, female. *Culex p. australicus*: c, male. *Tripteroides tasmaniensis*: d, male. *Aedes sagax*: e, male; f, female. *Aedeomayia venustipes*, g, male. *Aedes alternans*: h, female.

above the anterior spiracle the lateral margin forms a slight prominence called the scutal angle. Along the mid line of the scutum there is a row of acrostichal bristles and, on either side of these, a row of dorsocentral bristles. In addition, the scutum has a rather uniform clothing of scales which are often arranged in distinctive colour patterns; a median posterior area, known as the bare area, is devoid of scales and bristles. The paratergites are small narrow sclerites on either side of the scutum at about mid length.

The scutellum which lies immediately behind the scutum, being separated from it by only a narrow prescutellar area, is strongly convex; its posterior border is evenly rounded in Anophelines but in Culicines is trilobed and bears long marginal bristles; scales may be present on its upper surface. The postnotum is a convex sclerite and is usually bare.

The side walls of the mesothorax, the mesothoracic pleura, are made up of a number of sclerites and membranous areas. The membranous subspiracular area is situated below the posterior pronotum, and posterior to

STRUCTURE AND BIOLOGY OF THE ADULT

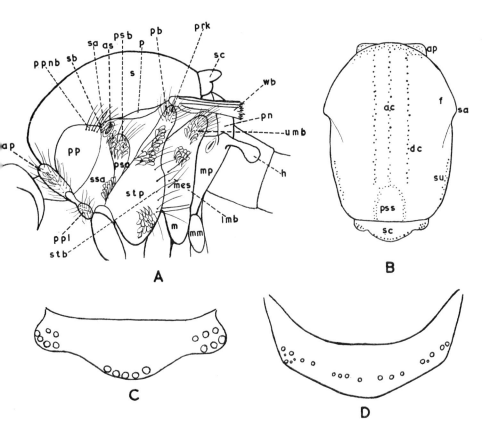

Fig. 4 Parts of thorax and pleural bristles. A, lateral view; B, dorsal view with positions of bristles indicated; C, scutellum of a Culicine; D, scutellum of an Anopheline. *ac*, acrostichal bristles; *ap*, anterior pronotum; *as*, anterior spiracle; *dc*, dorsocentral bristles; *f*, fossa; *h*, halter; *lmb*, lower mesepimeral bristles; *m*, meron; *mes*, mesepimeron; *mm*, metameron; *mp*, metapleuron; *p*, paratergite; *pb*, prealar bristles; *pn*, postnotum; *pp*, posterior pronotum; *ppl*, propleuron; *ppmb*, posterior pronotal bristles; *prk*, prealar knob; *psa*, postspiracular area; *sa*, scutal angle; *sb*, spiracular bristles; *sc*, scutellum; *ssa*, subspiracular area; *stb*, sternopleural bristles; *stp*, sternopleuron; *su*, supra-alar area; *umb*, mesepimeral bristles; *wb*, wing base.

it and to the anterior spiracle is the postspiracular area. Below these two areas is a large sclerite, the sternopleuron, which has a narrow upward extension, terminating in the prealar knob. The mesepimeron is adjacent to the sternopleuron; below it is the meron and above it the area of insertion of the wing.

The mesopleural bristles and scale patches are very important in the identification of some genera and many species. Of particular importance are: the presence or absence of postspiracular bristles, the number and distribution of bristles on the sternopleuron, the prealar area and prealar knob, and the presence of one or two series of bristles on the mesepimeron. An upper series situated near the upper posterior angle is usually present but lower mesipimeral bristles, situated near the anterior margin of the sclerite, may be lacking; this is an important character in *Aedes*. Taxonomic descriptions of mosquitoes usually include many references to the distribution of mesopleural scales.

Metathorax. This is much reduced and rarely used for purposes of identification.

WINGS (fig. 5)

The wing venation of mosquitoes is rather uniform but sometimes offers useful characters: the length of the upper fork cell relative to its stem and the relative positions of cross-vein r-m and base of M_{3+4} distinguish some species. The wing membrane is usually clear but may have darkened areas.

Scales are present on the veins and are of two types: the squame scales

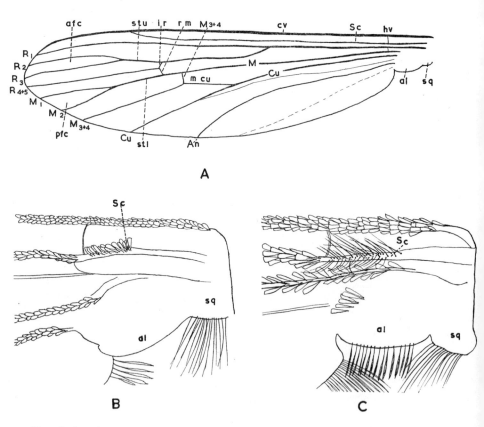

FIG. 5 A, wing of mosquito: *afc*, anterior forked cell; *al*, alula; *An*, anal vein; *Cu*, cubitus; *Cv*, costal vein; *i-r*, inter-radial cross-vein; *M*, media; *m-cu*, media-cubital cross-vein; *pfc*, posterior forked cell; *R*, radius; *r-m*, radio-median cross-vein; *sc*, subcostal vein; *sq*, squama; *stl*, stem of lower fork; *stu*, stem of upper fork. B, base of wing of *Anopheles*. C, base of wing of *Culiseta*: Sc, base of subcostal vein.

STRUCTURE AND BIOLOGY OF THE ADULT

which lie close to the veins and the plume scales which are narrower and semi-erect. Long narrow scales form a fringe from the apex of the wing to the alula, the more distal of the two small lobes at the base of the wing. In Anophelines the alula and the proximal lobe, the squame, are bare but in most Culicines, and in all Victorian species, they are fringed with hairs. The distribution, shape and colour of scales on the wing is of taxonomic importance, e.g. the absence of scales on the underside of the subcosta is a characteristic feature of *Culiseta*.

Halteres. The colour of the terminal knob and the presence or absence of scales on it sometimes provide specific characters.

LEGS

Each leg is composed of a coxa, trochanter, femur, tibia and a 5-segmented tarsus. The terminal tarsal joint bears a pair of claws which may be simple or toothed. In the female of many species, toothed claws occur only on the front and mid legs; in the male, the claws on the front legs and on the mid legs may be unequal in size and only the larger one may be toothed. Between the claws is an empodium (fig. 6) and below them, in *Culex* but not in other genera, is a pair of pad-like pulvilli. The legs may be clothed entirely with dark scales but often pale ones are present as well; these may be scattered or arranged in stripes or more commonly in rings, particularly on the tarsal segments.

ABDOMEN

The abdomen consists of ten segments, of which the two terminal ones are modified in relation to sexual functions. Each of the first eight segments consists of a dorsal tergite and a ventral sternite joined by membranous pleura. The tergites are collectively referred to as the dorsum, the sternites as the venter. In Culicines the abdomen is usually clothed with scales and their patterns on the tergites or sternites are often of diagnostic value. In Anophelines the abdomen is typically bare but may have a few scales.

The modified IXth and Xth segments form the terminalia or hypopy-

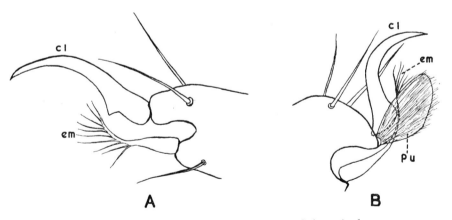

FIG. 6 Tip of a 5th tarsal segment of A, *Culiseta*; B, *Culex*. *cl*, claw; *em*, empodium; *pu*, puvillus.

gium. The terminalia of the female are rarely used in taxonomic descriptions of species, but often provide important generic and subgeneric traits. The male terminalia are of the greatest importance in taxonomy.

MALE TERMINALIA (fig. 7)

In the males of all mosquitoes the terminalia, together with the VIIIth

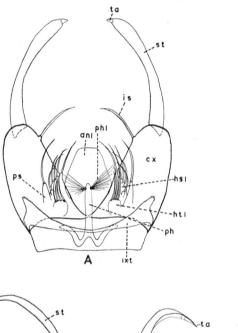

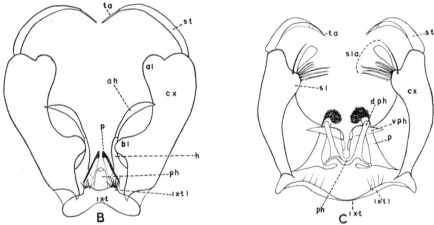

FIG. 7 Diagram of male terminalia. A, *Anopheles*; B, Aedes; C, Culex. *ah*, appendage of harpago; *al*, apical lobe; *anl*, anal lobe; *bl*, basal lobe; *cx*, coxite; *dph*, dorsal process of phallosome; *h*, harpago; *hsl*, sternal lobe of harpago; *htl*, tergal lobe of harpago; *is*, internal spine; *p*, paraproct (Xth sternite); *ps*, phallosome; *phl*, leaflets of phallosome; *ps*, parabasal spine; *sl*, subapical lobe; *sla*, appendages of subapical lobe; *st*, style; *ta*, terminal appendage of style; *vph*, ventral process of phallosome; IX-t, IXth tergite; IX-tl, lobe of IXth tergite.

segment, rotate through 180° so that the tergites of their segments come to occupy a ventral position and the sternites a dorsal position.

The terms tergite and sternite as applied to the terminalia always refer to their position before rotation and use of the terms dorsal and ventral is generally avoided. The terminalia comprise the following parts:

IXth tergite. This is a narrow sclerite, constricted in the middle and expanded laterally to form a pair of lobes which bear bristles. The shape of the tergites and the number of bristles are of importance in the identification of some species.

IXth sternite. As this offers few useful characters it is usually omitted from descriptions and drawings of the terminalia.

Coxites. The coxites are a pair of stout processes arising from the IXth segment. The inner surface of each bears one or two lobes variously armed with hairs and bristles. In many genera there is a basal lobe. It is most strongly developed in *Aedes* and in this genus there is a small associated structure, the harpago (or claspette), arising from the membrane joining the bases of the coxites. In *Anopheles* the lobe is represented by a group of parabasal spines and here also there are claspettes. In some species of *Aedes* there is a second lobe, the apical lobe. In *Culex* there is only a subapical lobe which probably represents the displaced basal lobe; it tends to be subdivided into two parts and bears highly specialized setae.

Style. This is a simple appendage articulated with the apical part of the coxite; it has a spine at, or near, its tip.

Xth segment. This is made up of a small tergite and a sternite consisting of two ventrolateral plates called paraprocts. The paraprocts are strongly developed in *Culex* and in this genus are armed with a cluster of spines at the tip; in other genera the spines are reduced in number and in *Aedes* are lacking.

Aedeagus. Under this term are included all the structures surrounding the opening of the genital duct. The central portion of the aedeagus is the phallosome (or mesosome), the structure of which provides important taxonomic characters. It may be in the form of a tube or scoop, as in *Ochlerotatus,* or a pair of simple lateral plates as in *Culiseta;* in *Culex* it may bear various arms and teeth.

Time of Activity

On the basis of their periods of activity, mosquitoes may be classified as diurnal, crepuscular or nocturnal. Activity, in this connection, refers particularly to biting activity since this is observed and measured more easily than such things as flight or mating.

Most Victorian species of *Aedes, Tripteroides, Mansonia* and *Culiseta* are diurnal or day-biting mosquitoes. This is also true of a few species of *Culex,* e.g. *C. orbostiensis* and of *Anopheles* (e.g. *An. atratipes*) but most of the species of these two genera and probably most zoophilic species of other genera remain in seclusion during the day, becoming active only after sunset. The above classification, however, is not a rigid one, for biting activity is influenced by meteorological factors. On hot summer days, species of *Ochlerotatus,* normally diurnal, become crepuscular and Lee *et al.* (1957) found that all the species they recorded, including some with well-defined morning peaks of biting activity, were active in the early evening. Diurnal activity was most intense at times of high atmospheric humidity and low illumination. These factors and also air turbulence influenced both the duration of the evening period of activity and its intensity; exceptional activity was recorded on calm evenings when a gentle convection current could be detected a few feet above ground level.

In Victoria no detailed observations have been made on daily cycles of other activities. Crepuscular and nocturnal species mate and lay their eggs at dusk or later but although it is stated (Bates, 1949) that, in general, day-biting mosquitoes also mate and lay eggs during the day, this seems to be exceptional among Victorian species. Only two, *Ae. postspiraculosis* and *Culiseta hilli*, have been observed to mate during the day and it seems probable that in the other diurnal species mating and oviposition occur during the evening period of activity.

Mating

It has long been known that male mosquitoes exhibit swarming behaviour and that mating occurs at this time. Females enter the swarm, the males seige them and copulation is completed as they leave the swarm or after settling on vegetation. However, the widely held view that swarming is part of the sexual behaviour is rejected by Nielsen and Haeger (1960). Their conclusion, based on a wealth of evidence, is that male swarming and mating are independent habits.

Swarming and mating habits vary even between closely related species. Thus in *C.p. australicus,* swarms form shortly after sunset in open country in the vicinity of breeding places; they consist of 100-150 males moving rhythmically in a vertical direction some 5-6 feet above ground. However, in the closely related *C.p. molestus,* the swarms, consisting of only 10-30 males, are found just above the water surface in tanks or barrels, or sometimes between buildings. *C. annulirostris* swarms after sunset; the swarms consist of several hundred males some 5-10 feet above the ground close to or almost in the tree canopy. *Culiseta hilli* swarms only occasionally; more usually males are seen during the day flying above the grass between shrubs in 'searching flight' and coupling with any active females.

Although, in nature, *C.p. molestus* may show swarm behaviour, it will mate with resting females in small cages (30-50 cubic inches). This faculty, known as stenogamy, is relatively rare and is known to occur in only six other Victorian mosquitoes: *C.p. fatigans, C. globocoxitus, T. tasmaniensis, Ae. postspiraculosis, Ae. australis* and *C. hilli*.

Mosquitoes that will not mate in small cages are described as eurygamous and it may be difficult, or impossible, to achieve natural mating of such species in the laboratory, e.g. in an experiment with *C.p. australicus* several hundred adults were liberated in a room some 500 cubic feet in size and although the temperature, light intensity and humidity were varied, not a single female was fertilized. In order to obtain full details of the life history of such a species it is necessary either to employ the forced-mating technique (McDaniel, 1957; Wheeler, 1962) or to feed wild-caught females which are almost invariably fertilized. Species which mate in the laboratory do so within 2-3 days of emergence from the pupa and this is probably true of all species in nature.

Feeding

As a general rule, mosquitoes will not lay eggs until they have had at least one blood meal. However, there are some species, described as autogeneous, in which this is not necessary for the production of the first batch of eggs, although a blood meal is a prerequisite for the laying of any subsequent batch.

Autogeny in *C.p. molestus* has been studied by many authors and most recently by Clements (1956) and Twohy and Rozeboom (1956). Their work showed that the larval fat body was larger in *C.p. molestus* than in the anautogenous *C.p. pipiens* and that the pupae and recently-emerged adult females of *C.p. molestus* contained larger amounts of lipids, glycogen and proteins. The larger reserves accumulated by *C.p. molestus* were not the result of better larval nutrition but were due to an inherent physiological mechanism. Clements (1956) also demonstrated that liberation of gonadotrophic hormone initiated autogenous ovarian development within a few hours of emergence of the adult.

The frequency of autogeny in wild populations of *C.p. molestus* in Victoria is high; it ranges from 46·6 to 94·2 per cent (Dobrotworsky, 1954b). Autogeny has been observed in only two other Victorian species: *Ae. australis* (Woodhill, 1936) and *T. tasmaniensis* (Dobrotworsky, 1954b). It is interesting to note that all these autogenous mosquitoes are also stenogamous. However, the converse is not true: both *Ae. postspiraculosis* and *C. hilli* are stenogamous but neither is autogenous. All anautogenous species require a blood meal to ensure maturation of the eggs. In *Culex*, a single meal is adequate for the production and oviposition of a full batch of eggs but this appears not to be the case with some species of *Mansonia* and *Aedes*. In the laboratory, *M. linealis, M. aurata, Ae. andersoni, Ae. silvestris* and *Ae. nivalis* laid no eggs until they had had a second blood meal. That this is not simply a laboratory phenomenon is suggested by the observation that in nature many species of *Aedes,* including those mentioned above, come to bite while still having undigested blood in the gut (Lee, 1957; Dobrotworsky, unpublished). A need for two blood meals has been demonstrated in *Anopheles gambiae* (Davidson, 1954); the first brings the oocytes to second-stage development, the second brings them to maturation.

Some mosquitoes are known to feed on the blood of reptiles and frogs. In Queensland *Uranotaenia albescens* Taylor and *Culex (Lophoceraomyia)* sp. have been recorded biting a large green frog (Marks, 1950), but this has not been recorded for any species in Victoria, which, as far as is known, attack only birds and mammals. The majority attack man but the following species are not known to do so: *An. stigmaticus, An. pseudostigmaticus, C. inconspicua, Aedomyia venustipes, C.p. australicus, C. globocoxitus, C. fergusoni.*

Most of the man-biting species attack all parts of the body indiscriminately but some are more selective; *Culiseta,* during the day, attacks the legs, while *Tripteroides, C. orbostiensis* and *Ae. dobrotworskyi* confine their biting mainly to the upper parts of the body, in particular, the face.

Very little is known about the feeding preferences of Victorian mosquitoes, although some information has been obtained by Lee *et al.* (1954), Reeves *et al.* (1959) and Lee *et al.* (1957). Feeding preferences have an obvious significance in relation to the transmission of disease organisms; that they are also important in relation to fecundity is indicated by observations on *C.p. australicus.* This species does not bite man but can be induced to do so in the laboratory. The resulting egg rafts contain only 30-73 eggs whereas those laid after feeding in nature contain 113-503 eggs (Dobrotworsky and Drummond, 1953). Feeding of the males and females on nectar from plants has been recorded in various parts of the world (West and Jenkins, 1951; Downes, 1958; Sandholm and Price, 1962), but no observations have been recorded in Victoria.

The sugar yields energy for flight and laboratory colonies are provided with it in order to ensure normal longevity. The nectar meal is also a source of water and this may be of critical importance in hot weather; during the summer the longevity of laboratory stocks of several local species is greatly increased if they have access to water.

After one or two blood meals the female retreats into dark and humid places such as dense grass or bushes, crevices and holes under tree roots, hollow tree stumps, animal burrows, etc. There the female remains inactive until the eggs are fully matured.

Oviposition

Mosquitoes lay their eggs either directly on the water surface or, as is the case with *Aedes, Tripteroides* and some species of *Culiseta,* on the moist rock, soil, or debris immediately above the water level. The larval stages are invariably aquatic but each species of mosquito has a more or less restricted range of breeding places and it seems clear that the ecological distribution of larvae is largely determined by the selection by adults of particular oviposition sites.

In some species the primary selection seems to be determined by the position of the water relative to ground level. Species of *Tripteroides* normally breed in tree holes; they will oviposit freely in water tanks and in glass jars placed in cavities in trees (or suspended in the branches) but they hardly ever breed in ground water. In contrast to this behaviour, many species of *Culiseta* show a strong preference for subterranean water and habitually breed in the burrows of land crayfish.

Laboratory investigations of oviposition are often unsatisfactory because of the abnormal behaviour of caged mosquitoes (Bates, 1949) but it has been shown that oviposition may be influenced by a variety of environmental factors. For some species, water temperature is important: for *Aedes rupestris* the preferred water temperature is 85°F, for *Ae. rubithorax* only about 59°F (Dobrotworsky, 1958). This links up nicely with the observation that *Ae. rupestris* breeds only in pools exposed to the sun whereas *Ae. rubrithorax* uses shaded pools.

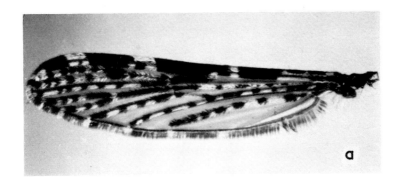

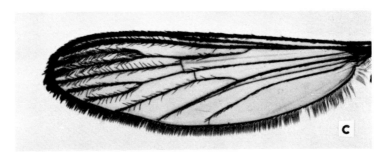
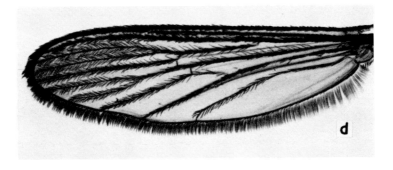

PLATE I Wings of mosquitoes. *a, Anopheles annulipes* Walker. *b, Aedeomayia venustipes* (Skuse). *c, Culex fergusoni* Taylor. *d, Culex pipiens molestus* Forskal.

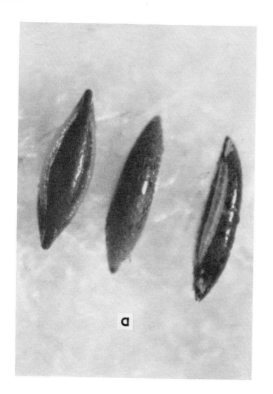

PLATE II   Eggs of mosquitoes. *a*, *Anopheles* sp., ×70. *b*, *Aedes nigrithorax* (Macquart), ×35. *c*, *Culiseta victoriensis* Dobrotworsky, ×35.

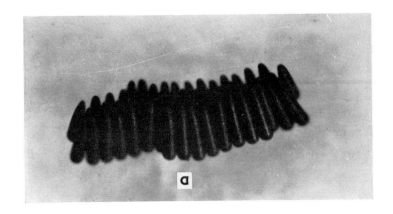

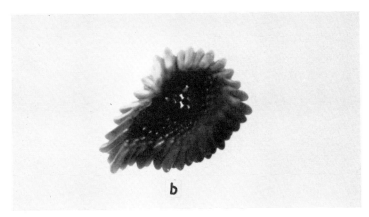

PLATE III Egg rafts of mosquitoes. *a, Mansonia aurata* Dobrotworsky, ×27. *b, Culiseta inconspicua* Lee, ×15. *c, Culex fergusoni* Taylor, ×27.

It is likely that water temperature and salinity are important for many species. An additional factor influencing oviposition by some, and possibly many, species of *Aedes* is the physical nature of the surface of the container (O'Gower, 1958). *Ae. australis,* in the Sydney area, breeds only in sandstone salt-water pools, but laboratory experiments have shown that it prefers fresh water for oviposition (Woodhill, 1941; O'Gower, 1958). This preference, however, is overridden by a stronger preference for a moist porous surface, particularly if it is rough. As O'Gower has pointed out, this behaviour can explain the occurrence of *Ae. australis* in salt-water pools but not its absence from fresh-water rock pools; the larva can tolerate a wide range of salinities and can be reared to maturity in tap water (Woodhill, 1936). It is of interest to note that in Victoria *Ae. australis* is not restricted to rock pools; it also breeds in salt-water ground pools.

Some species of *Aedes,* including *Ae. flavifrons,* do not lay their eggs on exposed surfaces but insert them into cracks or crevices in the soil or under debris. This, perhaps, is a response to a humidity gradient. Humidity is certainly a factor in the selection of a general site for oviposition (Knight and Baker, 1962). Some species have been observed ovipositing in the damp debris left in the bottom of a depression after the free water has evaporated.

Other factors known to influence oviposition include illumination and degree of pollution of the water.

2

STRUCTURE AND BIOLOGY OF THE IMMATURE STAGES

Eggs

The shell of the mosquito egg consists of a delicate vitelline membrane enclosing the yolk and of an outer chorion which consists of two layers, an inner endochorion and a much thinner and transparent exochorion. The egg when laid is white but within a few hours the endochorion hardens and, at the same time, darkens so that the egg becomes, in some species brown, in others black.

The surface of the endochorion has a sculptured appearance due to the presence of bosses of various sizes arranged in regular patterns which provide a means of identifying species, even closely related ones (Horsfall et al., 1952; Craig and Horsfall, 1960). These authors employed a quite elaborate technique (Craig, 1955) for recording these patterns but a simpler one has been devised by Pillai (1962a) who found that the surface markings of the exochorion are identical with those of the endochorion, and could be impressed directly on a celloidin film. Pillai's observations indicated that the sculpturing of eggs would enable identification of Australian species of *Aedes*.

The eggs of Anophelines are laid singly on the water surface and are readily recognized by the pair of lateral floats (fig. 8a, pl. IIa). These are formed by curved projecting folds of exochorion which also form a narrow frill extending around each end of the egg.

Of the Culicines, all species of *Culex*, all Victorian species of *Mansonia* and some species of *Culiseta* deposit their eggs in rafts floating on the water surface (pl. IIIa-c). The eggs, which are circular in section, are somewhat tapered at one end and bluntly rounded at the other. The eggs of *Culex* have a peculiar cup-like structure, the *corolla*, at the rounded anterior end. All species of *Aedes* and *Tripteroides* and some of *Culiseta* deposit their eggs singly (pl. IIb, c; fig. 8c, e, f, g, i, j).

In all species of mosquitoes, as far as is known, embryonic development is completed within a few days of laying and, in those which oviposit on the water surface, the larvae then emerge. However, in *Aedes*, *Tripteroides* and some species of *Culiseta* the eggs which are laid on soil, rock surfaces or debris above water level do not hatch until after they have been submerged as a result of rain or artificial flooding.

The factors promoting hatching are complex and not fully understood. In some species of *Aedes*, including all those in the subgenus *Finlaya*,

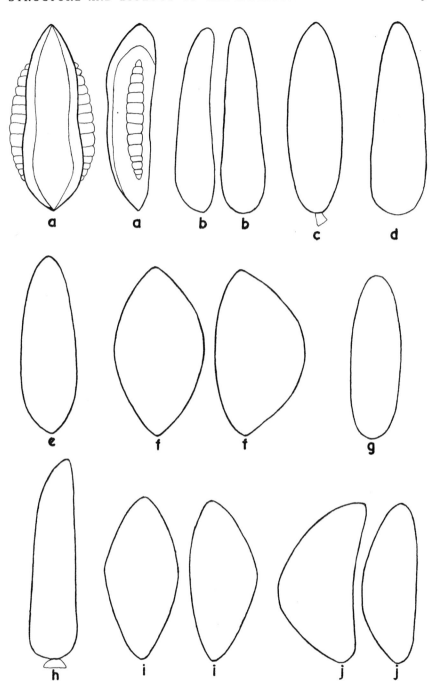

FIG. 8 Outlines of eggs of Victorian mosquitoes. *a*, *Anopheles annulipes*; *b*, *Culiseta inconspicua*; *c*, *Culiseta frenchii*; *d*, *Mansonia linealis*; *e*, *Aedes rubrithorax*; *f*, *Aedes alternans*; *g*, *Aedes dobrotworskyi*; *h*, *Culex p. australicus*; *i*, *Tripteroides tasmaniensis*; *j*, *Tripteroides atripes*. Not all to same scale. *a*, *b*, *f*, *i* and *j*, two views.

hatching is possible once embryonic development is complete. In nature, hatching may occur within a few days or only after an elapse of several weeks or months, but in some of these species, and possibly in all, submergence does not result in immediate hatching of all the eggs. Lee et al. (1957) have observed that in *Ae. rubrithorax* a proportion of viable eggs may remain unhatched after five weeks immersion; in nature they would probably have experienced several periods of exposure and immersion.

In other species of *Aedes,* belonging to the subgenus *Ochlerotatus,* the eggs enter a diapause which is terminated only by a more or less prolonged exposure to particular combinations of temperature and moisture (Pillai, 1962b). Following this 'conditioning', the eggs will respond to a hatching stimulus. In the laboratory, hatching of conditioned eggs can be induced by exposure to low oxygen tensions (Gjullin *et al.*, 1941), a stimulus which in nature would be provided by a dense growth of microorganisms. Other hatching stimuli used in laboratory experiments include a variety of enzymes and reducing agents (Rouband and Colas-Belcour, 1927). In these species, also, the hatching of some eggs may be delayed so that a complete hatch is not obtained until the sequence of treatments (conditioning, immersion, hatching stimulus) is repeated. It is possible that if given a sufficiently strong stimulus all the eggs would hatch within a short time but in the conditions prevailing in nature it is likely that hatching is usually spread over a longer period. Lee *et al.* (1957) have commented on the adaptive significance of delayed hatching. The temporary pools used as breeding sites by many of these species sometimes do not persist long enough to permit larval and pupal development to be completed. In such circumstances, delayed hatching could prevent local extinction; some viable eggs would remain to take advantage of later rains.

Delayed hatching of eggs implies that they have some resistance to desiccation but in this respect there is great variation between species. The eggs of *Ae. notoscriptus, Ae. imperfectus* and *Ae. rubrithorax* do not survive more than 1-2 weeks when held at a relative humidity of 53 per cent (Pillai, 1962b); Dobrotworsky (1958) found an exposure of 2 days at a relative humidity of 53-58 per cent caused almost total mortality of eggs of *Ae. rubrithorax* but only a 5 per cent mortality in the closely allied *Ae. rupestris.* The highest level of resistance occurs in species of *Ochlerotatus* from northern Victoria: the eggs of *Ae. sagax, Ae. theobaldi* and *Ae. vittiger* may survive 10-15 weeks exposure at a relative humidity of 53 per cent (Pillai, 1962b).

External Anatomy of Larva

There are four larval stages which differ in some degree from one another. The following account refers to the fourth-stage larva which is the one usually described in systematic work. The terminology is that of Belkin (1950, 1953).

HEAD (figs. 10, 11)

The head capsule is formed of three large sclerites, a dorsal clypeus and a pair of lateroventral epicranial plates bearing the eyes and antennae. The antenna is slender and bears six setae: seta 1, known as the antennal tuft,

STRUCTURE AND BIOLOGY OF THE IMMATURE STAGES 21

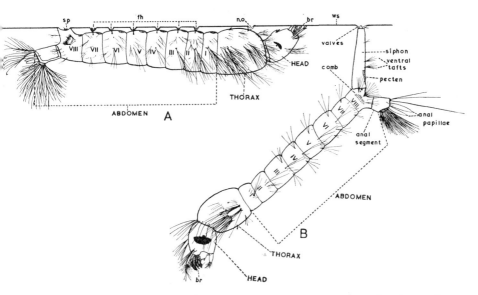

FIG. 9 Anopheline and Culicine larvae. A, lateral view of Anopheline larva; B, lateral view of Culicine larva. *br*, mouth brush; *fh*, float hairs; *no*, notched organ; *sp*, spiracle; *ws*, water surface.

varies in position and structure in different species; the remaining five setae may all be situated at the tip or setae 3 and 4 may be removed from it. The shaft of the antenna may be smooth or spiculate.

There are 10 pairs of dorsal head setae. Of these, setae 4, 5, 6 and 7 are of most importance in respect both of their structure and their positions relative to one another; the other setae are very small in Culicine larvae and, as a rule, are not included in descriptions of species of this subfamily.

The most conspicuous components of the mouth parts are the paired mouth or feeding brushes. Each brush consists of a group of rather stiff bristles arising ventrolaterally on the labrum. Posteriorly to the labrum are the mandibles, maxillae and labium; the anterior part of the labium, the mentum, is roughly triangular in shape with an apical tooth and a series of lateral teeth on each side. The number and size of the teeth of the mentum are of diagnostic value but the other mouth parts are rarely used in identification.

THORAX (figs. 10, 11)

The three thoracic segments are fused together and are defined only by the distribution of setae. The prothorax has fifteen pairs of setae, the mesothorax fourteen, and the metathorax thirteen. In Anopheline larvae, setae on all the thoracic segments are important for identification, but in Culicines prothoracic setae 1-7 are particularly useful and these alone will be mentioned in descriptions.

ABDOMEN (figs. 9, 12, 13)

The abdomen has nine segments. In Anopheline larvae each segment

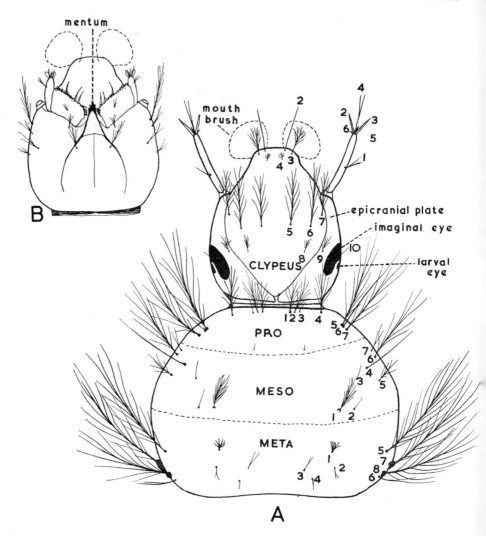

Fig. 10 Head and thorax of *Anopheles* larva.
A, dorsal view; B, ventral view of head.

has a dorsal sclerotized plate and in most species each of the IInd to VIIth segments has a pair of palmate setae; these structures are not present in Culicine larvae.

The structure of the VIIIth segment and the IXth (anal) segment is of great taxonomic importance as it provides both generic and specific characters. In Anophelines, the VIIIth segment bears the spiracular apparatus, a zone of spines called the pecten and five pairs of setae known as the pentad setae. The VIIIth segment of Culicine larvae bears the siphon and, on either side, pentad setae and a patch of scales, called the comb. At its distal end the siphon has five valve-plates surrounding the spiracular

STRUCTURE AND BIOLOGY OF THE IMMATURE STAGES

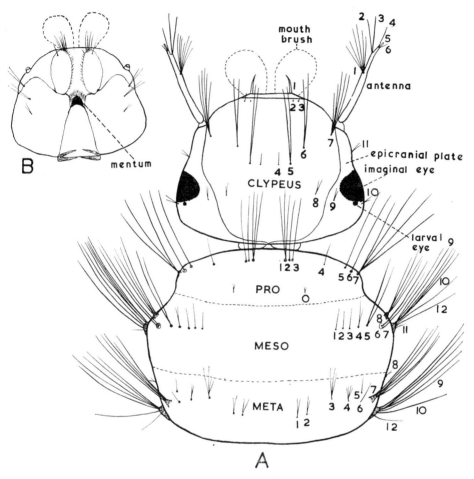

FIG. 11 Head and thorax of *Culex* larva. A, dorsal view; B, ventral view of head.

opening. The siphon index is of diagnostic importance; it is the ratio of the length of the siphon (excluding the valves and the acus) to its width at the base. Other taxonomically useful characters are the number and shape of the teeth of the pecten and the number, position and structure of the siphonal setae. There are nine pairs of these setae. Seta 1, in most Culicines, consists of one or more ventrolateral tufts. The remaining setae are situated distally; seta 2 arises from the apex of the siphon, setae 3-9 from the valve plates. However, setae 3, 4, 5 and 7 are minute and are not shown in the figures.

In the genus *Mansonia* the larval siphon is much modified in association with the habit of obtaining oxygen directly from aquatic plants. Beyond about mid length the siphon tapers sharply and is strongly sclerotized; it bears saw-like structures and apical teeth which are used in penetration of plant tissues and for attachment.

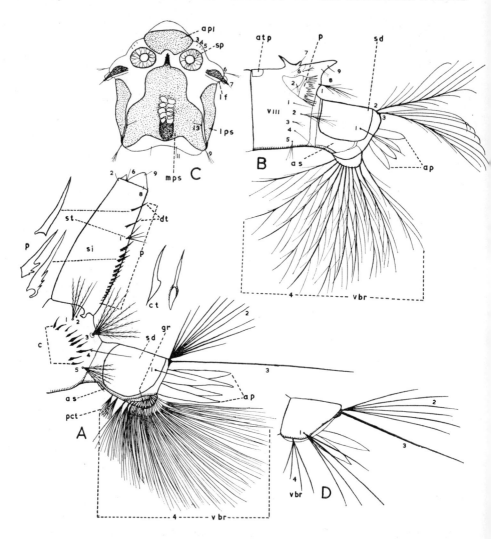

Fig. 12 Terminal segments of larva. A, *Aedes*; B, *Anopheles*; C, spiracular apparatus (dorsal view) of *Anopheles* larva; D, anal segment of *Tripteroides*. *ap*, anal papillae; *apl*, anterior plate; *atp*, anterior tergal plate; *c*, comb; *ct*, comb teeth; *dt*, detached teeth of pecten; *gr*, grid; *lf*, lateral flaps; *lps*, lateral plate of scoop; *mps*, median plate of scoop; *p*, pecten; *pct*, precratal tufts; *sd*, saddle; *si*, siphon; *sp*, spiracle; *st*, siphonal seta; *vbr*, ventral brush. The setae are numbered.

The main features of the IXth or anal segment are the saddle, the setae and the anal papillae. The saddle is a dorsal sclerotized plate which partly, or sometimes completely, encircles the segment; its surface may bear minute denticles. The most prominent setae are those forming the ventral brush (seta 4). This consists of a row of hair tufts the bases of which may fuse together to form a grid; the tufts arising from the grid are called the

STRUCTURE AND BIOLOGY OF THE IMMATURE STAGES

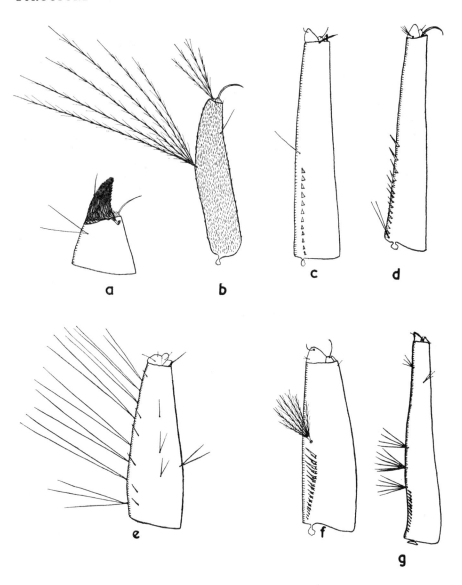

Fig. 13 Siphons of fourth-stage larva. *a, Mansonia*; *b, Aedeomyia*; *c, Culiseta littleri*; *d, Culiseta victoriensis*; *e, Tripteroides*; *f, Aedes*; *g, Culex*.

cratal tufts, those situated proximally to it, the precratal tufts. Seta 2 forms a smaller dorsal brush; setae 1 and 3 are usually simple.

The four anal papillae arise at the tip of the IXth segment. The length of the papillae varies in different species but since it is affected by physico-chemical factors in the water it may show considerable variations in larvae of the one species taken from different localities.

Biology of the Larva

Two methods of feeding are used by mosquito larvae. In one, rapid movements of the mouth brushes associated with the labium create water currents which draw floating particles towards the mouth; the particles are filtered out and swallowed. In the other method, food particles are scraped from the surface of water plants, dead leaves and other submerged objects by means of modified serrated hairs of the mouth brushes.

Anopheline larvae feed exclusively by the first method and mainly while at the water surface. Because they lack a siphon the larvae, when at the surface, lie in a horizontal position with the dorsal surface uppermost; efficient collection of food particles from the surface film is made possible by the extreme mobility of the neck which permits the head to be rotated through 180°.

The larvae of *Culex, Culiseta* and *Mansonia* are filter-feeders. This is also true of some species of *Aedes* but other members of this genus and all *Tripteroides* larvae feed mostly by scraping particles from submerged surfaces.

Culicine larvae feed at various depths but those that employ the filtering method are able to feed at the surface. Normally they hang head downwards with only the tip of the siphon breaking the surface but they are able to bend the body to bring the head to the surface and collect food particles while slowly rotating about the tip of the siphon, which acts as a pivot. In the great majority of mosquitoes the larvae are microphagous but in a few they are predaceous. The only Victorian species strictly of this habit is *Ae. alternans;* the second and later stages are entirely predaceous, feeding mainly on other mosquito larvae. However, *T. tasmaniensis* in overcrowded conditions may combine cannibalism with the usual microphagy; the larger larvae attack and eat the smaller ones.

The larvae, and also the pupae, of most species of mosquitoes react to passing shadows or to mechanical disturbances by moving to the bottom of the pool. The only larvae known not to show this alarm reaction are those of species of *Culiseta* which breed in underground water; they do not respond to passing shadows, and mechanical disturbances often result in their moving towards the surface.

In three species, *Ae. theobaldi, Ae. bancroftianus* and *Ae. postspiraculosis,* the larvae visit the surface only at long intervals. For the most part they are found either lying upside down on the bottom or hanging from submerged objects to which they attach themselves by means of a modified hook-like seta (seta 9) on the siphon; it is interesting that *Mansonia* larvae use a pair of modified siphonal setae for attachment prior to forcing the tip of the siphon into the tissues of plants. In these species of *Aedes,* the respiratory system, apart from a narrowing of the siphonal tracheae, is not structurally modified.

The duration of the larval stage varies from species to species but is always greatly influenced by water temperature. Broadly it can be said that, at the temperature prevailing in northern Victoria during the summer, larval development is complete in about 14 days. This would also apply to southern species breeding in exposed pools but if the winter is passed in the larval stage, or if the larvae live underground, full development may require several months.

Tolerance of high water temperatures varies greatly. Larvae of *Ae. rupestris* have been collected from pools with a midday water temperature of 100°F (Marks, Field Report Card), At the other extreme, *Culiseta* larvae from underground water cannot survive in water at 68°F. This temperature would be higher than any expected in their natural habitat for the water temperature in crayfish burrows does not exceed 60°F even in midsummer.

Larval Breeding Sites

Mosquitoes use as breeding sites almost all terrestrial accumulations of water. As already pointed out, larval habitat is determined mainly by the oviposition behaviour of the female but, since this is poorly understood, classifications of larval breeding sites are at present based largely on convenience.

The great majority of mosquitoes in Victoria breed in fresh ground water, a phrase that covers a wide variety of habitats differing both in size and in degree of permanence. At one extreme are permanent lakes and large swamps, at the other are small depressions such as hoof-prints which fill with water only after rain and retain it for relatively short periods. A factor, ignored in the classification adopted here, but one that is often important, is illumination: a swamp exposed to the sun, for example, is quite a different habitat from a swamp heavily shaded by trees.

There are, however, some species which rarely, if ever, breed in ground water. *Ae. dobrotworskyi,* as far as it is known, breeds only in the leaf axils of sword grass. *Ae. notoscriptus* and species of *Tripteroides,* in nature, are tree-hole breeders and though they also breed in a variety of artificial containers it is quite exceptional to find their larvae in ground water.

In the following classification, the number of different habitats has been exaggerated by making a broad distinction between natural habitats and those created by man, but this will serve to emphasize how human activities multiply breeding sites for mosquitoes. A particular example is the occurrence of enormous mosquito populations in irrigated areas along the Murray Valley; another is the abundance of *C.p. molestus* in many townships and holiday resorts in southern Victoria where almost the only breeding sites are septic tanks.

CLASSIFICATION OF BREEDING PLACES
I. Fresh Water
  A. Water at Ground Level or Below
    a. Natural Habitats
      1. Lakes—vegetated margins
      2. Streams—back waters and, if slow flowing, vegetated margins
      3. Swamps
      4. Temporary and semi-permanent ground pools, including hoof prints

        5. Rock pools—at sides of stream bed
        6. Flooded animal burrows
        7. Pits under uprooted trees
        8. Tunnels and underground pools of land crayfish
     b. Man-made Habitats
        1. Irrigation ditches and their overflows
        2. Small dams
        3. Excavations—pits, roadside ditches, surface gold diggings, shallow wells
        4. Wheel ruts
  B. Water Above Ground Level
     a. Natural Habitats
        1. Leaf-axils
        2. Tree holes
     b. Man-made Habitats
        1. Tanks, horse drinking-troughs, water butts, tins, discarded tyres
II. Water Heavily Polluted with Organic Matter
        1. Pools in rubbish dumps
        2. Drains and drainage pits
        3. Septic tanks
III. Brackish Water
        1. Marshes
IV. Salt Water
        1. Coastal rock and ground pools

Pupa (fig. 14)

The body of the pupa consists of two parts: an enlarged anterior part, the cephalothorax, and a narrow segmented abdomen terminating in a pair of paddles. No detailed account of the chaetotaxy of the pupa will be given here for although it can be used for the specific identification of pupae this is rarely necessary; nearly always larvae or adults are available and their identification is much easier.

The respiratory trumpets in Anophelines are short with a wide oblique opening; in Culicines they are usually either broadly conical or narrow and elongate, but the opening is rarely oblique. In *Mansonia* the trumpet is slender and rather rigid with a pointed tip which can be inserted into the tissues of aquatic plants.

The paddles are borne on the VIIIth segment of the abdomen. The setae on the posterior margin have diagnostic value.

The duration of the pupal stage depends upon the temperature of the water. In northern species such as *Ae. vittiger, Ae. vigilax* and *C. annulirostris* it may be as short as 2-3 days but in the species of *Culiseta* breeding in subterranean water at temperatures of 56-59°F it may last 15-17 days.

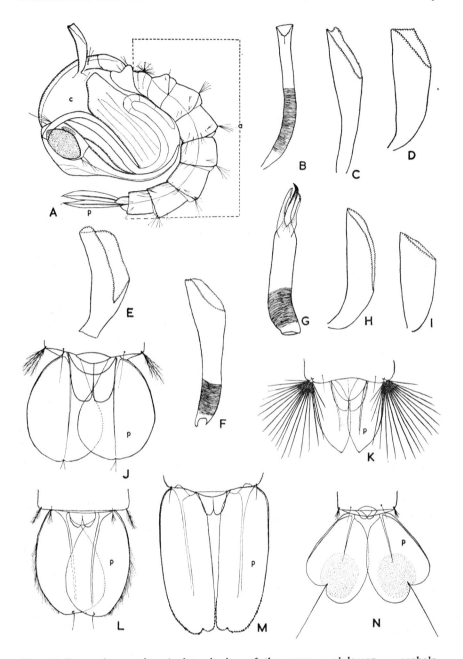

FIG. 14 Pupa of mosquito. A, lateral view of the pupa; *a*, abdomen; *c*, cephalothorax; *p*, paddle. B-I, respiratory trumpets: B, *Culex orbostiensis*; C, *Culex p. molestus*; D, *Aedes rubrithorax*; E, *Anopheles stigmaticus*; F, *Aedeomyia venustipes*; G, *Mansonia uniformis*; H, *Culiseta drummondi*; I, *Tripteroides atripes*. J-N, part of terminal abdominal segment and paddle: J, *Aedes purpuriventris*; K, *Tripteroides atripes*; L, *Anopheles stigmaticus*; M, *Mansonia uniformis*; N, *Aedeomyia venustipes*.

3

DISTRIBUTION AND COMPOSITION OF THE MOSQUITO FAUNA

In Australia, as in other parts of the world, the number of genera and species of mosquitoes reaches its maximum in the tropical and subtropical zones and decreases in the temperate regions. Thus in Queensland 11 genera and 150 species have been recorded whereas in Victoria there are only 7 genera and some 70 species. Nevertheless, species that penetrate into cooler regions may build up to huge populations.

Climate is undoubtedly a major factor influencing the distribution of mosquitoes, and in Victoria the most important climate barrier is formed by the Central Highlands. These are divided into the Eastern Highlands and the Western Highlands. The Western Highlands are of low elevation rarely rising to 3,000 feet but in the Eastern Highlands there are several peaks above 6,000 feet. In the alpine areas and high plains the mean minimum daily temperature falls below freezing point for five months of the year; the average annual rainfall is between 50 and 60 inches.

To the north of the Highlands lie the Murray Basin plains, and to the south the Gippsland and Western District plains. Further south are the southern uplands, the most important elements of which are the South Gippsland and the Otway Ranges; both are high-rainfall areas.

The Central Highlands form a boundary for summer heat and for rainfall between the northern and southern plains. The state has a regular weather cycle involving alternating invasions of air from the south and from the north. During the summer, cool southerly winds prevail and the period from one cool change to the next is commonly of 4 or 5 days. Many of the cold changes do not extend beyond the Highlands and hence the southern part of the state has a mild summer and also a higher annual rainfall. The average maximum daily temperature in January, the hottest month, is 80°F or less; the average annual rainfall is 25-30 inches.

North of the Highlands the average maximum daily temperature during January is 85°F or more; in the Mallee, the extreme north-west corner of the state, it is close to 90°F. Throughout the area the average annual rainfall is less than 20 inches; in the Mallee it is less than 15 inches.

The contrast between the hot north and the cooler wetter south is further illustrated by the length of the growing season for agricultural crops; in the north it is from 5 to 7 months; in the south, from 9 to 12 months.

The Central Highlands which form the boundary between these two regions also determine the boundary between the areas of distribution of the two major elements of the mosquito fauna.

The actual boundary is just to the north of the Highlands and is indicated by the 20-inch isohyet. This, of course, is not an absolute boundary; some numbers of the northern element penetrate into the southern part of the state and vice versa.

Within the northern element three groups can be recognized. The largest one consists of species which are either absent from the southern part of the state or occur there only sparsely. The most typical members are: *Ae. vigilax, Ae. theobaldi, Ae. alternans, Ae. bancroftianus, An. annulipes, T. atripes* and *C. annulirostris.* Another group including *Ae. multiplex, Ae. procax, Ae. subbasalis, C. postspiraculosis, C. pseudomelanoconia, C. orbostiensis, M. variegata* and *M. aurata* is mainly restricted to the south-eastern part of East Gippsland, an area which is climatically distinct from the rest of Victoria (Patton, 1955); it is characterized by relatively high minimal temperatures throughout the year and by a prolonged summer.

A third group of northern species has penetrated into the cooler southern part of the state. Some of these are confined to habitats similar to those occupied in warmer regions. *Ae. rupestris,* for example, breeds only in rock pools exposed to the sun in which water temperatures during the summer may rise to 90°F or higher. Others, such as *An. stigmaticus, T. tasmaniensis* and *Ae. rubrithorax,* have become fully adapted to a cool climate and share breeding sites with typical members of the southern element.

This southern element is distributed mainly on and south of the Central Highlands, that is, in the part of the state with a cooler climate and non-seasonal rainfall. Typical representatives are members of the genus *Culiseta, Culex fergusoni,* and species of the subgenus Ochlerotatus, e.g. *Ae. luteifemur, Ae. nivalis, Ae. silvestris, Ae. andersoni* and *Ae. purpuriventris.* Some members of the southern element, however, have spread, to varying degrees, into the drier and warmer areas north of the Highlands. Thus *Ae. camptorhynchus* is abundant on the southern coast but is also common in the Mildura area; *Ae. sagax* is found for the most part north of the Central Highlands and *Ae. vittiger* has a distribution which is essentially that of a typical member of the northern element.

The occurrence of two elements in the Australian fauna was first recognized by Edwards (1924) in relation to species of the subgenus *Ochlerotatus.* In one group he placed species possessing lower mesepimeral bristles and stated that they 'show little or no divergence in structure from the Holarctic forms'. The members of the other group are characterized by the absence of lower mesepimeral bristles and by having male terminalia similar to those of South American tropical species.

Mackerras (1927) supported Edwards' conclusion and referred to the first group as part of an 'Antarctic Group'. Later (1950) he adopted the older term 'Bassian element' and listed four characteristics of this element in the Diptera: (1) its members are relatively primitive, (2) all are present in Australia and in southern South America; some are also present in New

Zealand, (3) in Australia they occur predominantly in the south and (4) in the south and in high country the adults are on the wing in the summer; on the coast, further north, they are active in the spring.

Coupled with Mackerras's emphasis on the South American relationship of the Bassian element is his advocacy of a southern entry to Australia via an Antarctic land mass. This general concept, once widely accepted, is now rejected by many zoogeographers in favour of entry by a northern route (Abbie, 1941; Darlington, 1953) but it is still supported by some (Evans, 1959; Paramonov, 1959).

In contrast to Mackerras's view of the nature of the Bassian element, Key (1959) points out that, in relation to grasshoppers, the zoogeographical regions of Australia as redefined by Serventy and Whittell (1948) correspond closely with regions defined by climatic factors of fundamental importance in grasshopper ecology. In his view, the Bassian fauna is simply the fauna adapted to the climate of the Bassian region; the existence of widely different faunas in different regions of Australia does not necessitate an assumption of different external origins (Key, 1959).

However, if the term 'Bassian element' is to be used in relation to mosquitoes it should be applied to what has been referred to above as the southern element and this does not fully meet the criteria of either Mackerras or Key. The genus *Culiseta* is a typical component of the southern element. On morphological grounds, Edwards (1932) regarded it as one of the 'nearest living representatives of the primitive stock which has given rise to all Culicine mosquitoes'. Its distribution in Australia, nine species in southern Victoria, two in New South Wales, one in southern Queensland, two in South Australia, one in southern Western Australia and three in Tasmania, is typically Bassian but the genus is totally absent from South America.

The species of *Ochlerotatus* comprising Edwards' Group G (1932) are not primitive mosquitoes; in fact, the complexity and diversity of structure of the male terminalia suggest that the whole subgenus is a specialized one. However, it is probable that Group G represents an early phase in the evolution of the subgenus. The group occurs in southern South America but in Australia is not confined to the Bassian region; some members of it extend widely into the Torresian or Eyrean Regions (e.g. *Ae. vittiger, Ae. perkinsi, Ae. sagax*). The third component of the southern element is *C. fergusoni*. The genus *Culex* is specialized in some respects but Edwards (1932) believed that it evolved relatively early. *C. fergusoni* is a member of the *C. apicalis* complex (Dobrotworsky, 1956) which is not represented in South America.

These three components of the southern element have in common the fact that related groups of species are widely distributed in the Holarctic region, rare in the tropics and absent from the Oriental Region. It seems probable that *Culiseta* and the *Culex apicalis* complex evolved in the tropics and subsequently spread to the northern and southern temperate regions, being progressively displaced away from the tropics by later-evolving elements. Belkin (1962) believes that new major types of mosquitoes evolved in the tropics and principally in two areas: the Indo-

Pacific and the American Mediterranean between the two American continents.

Thus a definition of the Bassian element of the Australian mosquito fauna should not specify a South American relationship. This element is conceived as consisting of an ancient element of the fauna, with Holarctic affinities, primarily adapted to the cool climate and non-seasonal rainfall of south-eastern Australia but embracing species now adapted to hotter, and in some cases, arid conditions.

The conclusion that *Culiseta* and the *Culex apicalis* complex entered Australia from the north does not preclude the possibility of a southern entry via an Antarctic land mass for other faunal groups, but from this conclusion it does follow that such a route need not be invoked to account for the existence of a Bassian element in the Australian fauna.

Little need be said about the origin of the northern element of the Victorian mosquito fauna. It is part of the Indo-Malayan element which includes a great assemblage of relatively recent successful insects which entered Australia from the north during the Pleistocene, dominate the Torresian province and overflow quite widely beyond its limits (Mackerras, 1950).

Mention should be made of two recent introductions, *C.p. fatigans* and *C.p. molestus*. The first, a tropical form, was probably introduced with the first white settlers (Mackerras, 1950); *C.p. molestus* was perhaps introduced during World War II (Drummond, 1951).

Seasonal Abundance

In Victoria, only a few species of mosquitoes maintain reproductive activity during the winter; the great majority spend this period of the year either as larvae or as hibernating adults; all species attain their greatest abundance during the spring or summer. The pattern of seasonal changes in abundance is influenced to some extent by oviposition habits. Species that lay eggs directly on the water surface are invariably multivoltine and maintain a high population continuously during the warmer months; other species may be multivoltine but are potentially univoltine. In the northern part of the state, the first species to be abundant in the spring are *C. globocoxitus, An. annulipes* and *Ae. camptorhynchus*, all of which maintain reproductive activity at a low level during the winter. The population of *C. globocoxitus* builds up to a peak in the spring but declines during the summer. The other two species, however, maintain high numbers through to the autumn. *C.p. australicus*, which hibernates in the adult stage, shows the same pattern as *An. annulipes* but does not reach peak numbers until midsummer.

In the north all species of *Aedes* apparently overwinter as larvae which hatch in the autumn and early winter and pupate in the spring. The earliest species are *Ae. sagax, Ae. theobaldi* and *Ae. bancroftianus*. In natural conditions in the north these species may be univoltine but in irrigation areas further generations are possible.

The effect of irrigation practices are illustrated by the observation of Pillai (1962b) on *Ae. vittiger*. At Nyah, in northern Victoria, irrigation channels are flooded three or four times between December and April. The water overflows into shallow depressions which form breeding sites for *Ae. vittiger*. Within a few days eggs, laid previously, hatch and at the prevailing temperatures development is completed and another generation of eggs laid before the next flooding. The cycle is repeated so that four or five generations are possible annually. Whether *Ae. sagax* and *Ae. theobaldi* follow the same pattern is not clear; neither of these species is at all abundant during the late summer, even in irrigation districts.

In southern Victoria, the adult mosquito population remains high from spring until the end of summer or early autumn. The spring population is composed mainly of species of *Aedes*, including all species of *Ochlerotatus* and of *Pseudoskusea* and some species of *Finlaya*, e.g. *Ae. alboannulatus*. These species pass the winter as larvae in pools more or less exposed to the sun and are able to complete their development and pupate in early spring. Several generations may be completed before the adults disappear in midsummer.

A second group of spring mosquitoes includes all species of *Culiseta* and *Tripteroides*, and some species of *Finlaya* and *Culex fergusoni*. These usually breed in shaded pools, crayfish tunnels or tree holes and, because of the low water temperature, larval development during the winter is slow. These species become abundant later than those that breed in exposed pools.

A few species, including *C.p. australicus* and *C. douglasi*, hibernate as adults. They resume activity in the spring but the population builds up rather slowly and does not attain its peak until midsummer.

4

MOSQUITOES AS VECTORS OF DISEASE IN VICTORIA

In Victoria, there are very few diseases known to be carried by mosquitoes.

The bird-pox viruses. Seven viruses having the general characteristics of bird-pox viruses have been recovered from mosquitoes collected on both sides of the Murray River in the Mildura area during the summer of 1951-2. These viruses were isolated from the *C. pipiens* group and *C. annulirostris* (Reeves *et al.,* 1954) but since their transmission is believed to be mechanical and not to involve a biological cycle in the mosquito, it is probable that all those species which usually feed on wild or domestic birds should be regarded as vectors of the bird-pox viruses.

Myxomatosis. The history of the introduction of myxoma virus into Australia for the control of rabbits has been reviewed by Ratcliffe *et al.* (1952). The first field trials, made in 1933, were not encouraging but this was undoubtedly due to the fact that, for reasons of safety, the trials were confined to semi-arid pastoral zones where the only transmitting agents were fleas. Following World War II the project was revived and in the spring of 1950 virus was liberated at seven sites in the Murray Valley. After a slow start, major epidemics occurred and it became evident that mosquitoes were mainly responsible for the rapid spread of the disease. Mykytowycz demonstrated that the widely spread *C. annulirostris* was a vector on the Murray River flats and later showed that various arthropod parasites of rabbits could also function as vectors (Ratcliffe *et al.,* 1952). It became apparent that transmission of the virus was mechanical, a conclusion confirmed by Fenner, Day and Woodroofe (1952) and Day, Fenner and Woodroofe (1956), so that any species of mosquito, or other arthropod, that fed on rabbits was a potential vector.

A number of species of mosquitoes are now known to feed on rabbits with some regularity; their relative importance as vectors will vary in different areas and in different seasons. In Victoria the most important vectors are *An. annulipes, C. annulirostris* and *C.p. australicus* (from all of which virus has been recovered in the field), while there is strong presumptive evidence that *Ae. camptorhynchus, Ae. nigrithorax* and *Culiseta* spp. are locally and at times of some importance (Douglas, 1958). In New South Wales, *Ae. alboannulatus* is known to play a part in the transmission of myxomatosis, and it may safely be assumed to do so where it is prevalent in Victoria.

Murray Valley Encephalitis (M.V.E.). This disease has been recorded on several occasions in the north-western part of the state, mainly in the Mildura area; the most recent outbreak of severe human encephalitis occurred early in 1951, and a few cases occurred early in 1956. Mosquitoes were suspected of being the vectors but were not definitely implicated. Examination of a small sample (550 mosquitoes of 4 species) collected in April 1951, and a much larger sample (17,833 mosquitoes of 10 species) collected during the summer of 1951-2, failed to isolate any M.V.E. virus (Reeves *et al.*, 1954). Tests made on mosquitoes collected in 1956 were likewise negative (unpublished data).

Laboratory studies by McLean (1953) revealed that the eleven species of Culicines tested could all be infected with M.V.E. virus and, from 7 to 41 days later, could transfer the infection to chickens. The virus undergoes a biological cycle in these mosquitoes; during the first day after feeding on the blood of an infected chicken, the virus can be found in the abdomen of the mosquito but on the second day cannot be detected. On the third day it is present in increased quantity in all parts of the body including the salivary glands. However, when *An. annulipes* was similarly tested, the virus was found only during the first day after feeding and McLean concluded that this mosquito is incapable of transmitting M.V.E. virus under experimental conditions. More recently M.V.E. virus has been isolated from mosquitoes collected in northern Queensland (Doherty *et al.*, 1961, 1963). From 25,901 mosquitoes, representing 32 different species, 11 viruses were isolated. M.V.E. virus was isolated from *C. annulirostris* and *Ae. vigilax*. The discovery of the virus in *C. annulirostris* is of particular interest. This is common in the Murray Valley where, on ecological grounds, it had been regarded as the most important of the potential vectors of M.V.E. (Reeves *et al.*, 1954).

Malaria. Malaria is transmitted exclusively by anopheline mosquitoes of which there are four species in Victoria. Two of these attack man but *An. atratipes*, because of its relative rarity, can be ignored as a potential vector. *An. annulipes*, on the other hand, is widely distributed in Australia and is particularly abundant in warmer areas, including northern Victoria.

Roberts (1943) has shown that *An. annulipes*, in experimental conditions, is as hospitable to *P. vivax* and *P. falciparum* as is the notorious malarial vector, *An. farauti* Laveran. But, as is well known, a capacity to harbour the malarial parasite does not necessarily mean that a mosquito is an important vector in nature. Its importance will be influenced by its feeding preferences, the frequency with which it comes into contact with man (especially indoors), the abundance of the species, and its longevity.

An. annulipes certainly attacks man but as Lee *et al.* (1957) have observed, it also attacks cattle and rabbits and, indeed, shows a preference for bovine blood; this preference was also evident during M.V.E. investigation at Mildura. It rarely enters houses and, still more rarely, bites indoors. *An. annulipes* has never been incriminated as a malarial vector in Australia and no natural infection has been recorded (Lee and Woodhill, 1944). However, Lee and Woodhill suggest that, in exceptional circumstances, it could be responsible for brief epidemics of malaria.

PART II

Taxonomic

FAMILY CULICIDAE

The family Culicidae as defined by Edwards (1932) included three subfamilies, the Dixinae, the Chaoborinae, and the Culicinae, but following Stone (1956) it is proposed here to treat each of these as a full family. The family Culicidae is thus restricted to include only the mosquitoes. These are readily recognized, in the adult stage by the elongation of the mouthpaths to form a proboscis and the presence of scales on the wing veins, in the larval stage by the broad unsegmented thorax and simple antennae and, in the pupal stage by the moveable paddles and the open respiratory trumpets with the spiracle at the base.

The family Culicidae is divided into three subfamilies, the Anophelinae, the Culicinae and the Toxorhynchitinae, but the last of the three does not occur in Victoria.

The principal diagnostic features of the Anophelinae and Culicinae are set out below.

ANOPHELINAE
  Eggs   —Laid singly; lateral floats present.
  Larvae —VIIIth abdominal segment without respiratory siphon.
  Adults —Abdomen with few, if any, scales.
         Female palps about as long as proboscis.
         Male palps strongly clubbed.

CULICINAE
  Eggs   —Laid singly or in rafts; never with lateral floats.
  Larvae —VIIIth abdominal segment with elongate respiratory siphon.
  Adults —Abdomen densely clothed with scales, at least ventrally. Female palps usually less than one-quarter the length of the proboscis, rarely two-thirds its length.
         Male palps not clubbed, except *Ae. australis* (Erichson).

SUBFAMILY ANOPHELINAE

Anopheline mosquitoes are widely distributed in tropical and temperate parts of the world but the majority of species are confined to the warmer areas; they are always most numerous in flat, swampy country and relatively few species penetrate into highland forests. Of the four species of *Anopheles* known from Victoria only one, *An. annulipes,* is at all common. It is adapted to a variety of conditions and occurs through the state. Two others, *An. stigmaticus* and *An. pseudostigmaticus,* particularly the former, are adapted to cooler conditions and are found in the highlands. The fourth species, *An. atratipes,* is restricted to a narrow coastal zone where rainfall and humidity are high.

Characters of the Subfamily

ADULT

Clypeus longer than broad, rounded anteriorly. Palps in both sexes usually about as long as proboscis, except in the genus *Bironella*. Scutellum evenly rounded, except in *Chagasia*. Abdomen without scales or with a few loosely applied scales; sternites usually bare.

Male terminalia. Coxites short, usually without distinct basal lobe; style long, slender with short terminal appendage. Phallosome tubular, usually with reflexed appendage (leaflets) at tip. Female with one spermatheca. Legs very long and slender; no distinct tibial bristles; no pulvilli. Wings usually with distinct markings.

LARVA

Head usually longer than broad, freely rotatable. Head seta 2 close to front margin; 5, 6 and 7, pinnate, in a row across middle. Mentum rather long and narrow. Most of larger body setae pinnately branched. Abdomen with a series of palmate tufts. Siphon absent. VIIIth abdominal segment with lateral chitinous plate with pecten. Seta 2 and seta 4 (ventral brush) of anal segment irregularly branched.

EGGS

Eggs pointed at both ends with dorsolateral or lateral floats.

Anopheles is the only genus found in Victoria. Three Victorian species belong to the subgenus *Anopheles* which has a world-wide distribution and one species to the subgenus *Cellia* which is confined mainly to the Oriental, Ethiopian and Australian Regions, but also includes a few Mediterranean species.

Genus ANOPHELES Meigen

Anopheles Meigen, 1818, *Syst. Beschr. zweifl. Ins.*, 1: 10.
For synonyms see Edwards, 1932.

KEY TO VICTORIAN SPECIES OF THE GENUS ANOPHELES

Adults

1   Wings and legs profusely marked with white scales .. ... *annulipes*
    Legs dark scaled 2
2(1) Wings with some silvery patches of scales; fringe with distinct white spot at tip of wing. Legs not marked with white scales .. *atratipes*
    Wings entirely dark scaled. Basal four-fifths of hind femur creamy scaled 3
3(2) Only one upper sternopleural bristle. Knob of haltere pale scaled
    .. .. *pseudostigmaticus*
    Two or more upper sternopleural bristles. Knob of haltere black scaled .. .. *stigmaticus*

Subgenus ANOPHELES Meigen

Anopheles Meigen 1818, *Syst. Beschr. zweifl. Ins.*, 1: 10.
For synonyms see Stone, Knight and Starke (1959).

Characters of Subgenus

ADULT
  Propleural hairs usually numerous. Spiracular hairs usually present and rather long. Prealar hairs present. Wings usually dark; if with pale markings, the bases of fork-cells and areas of veins immediately adjacent to cross-veins almost always dark. Male terminalia with 1-3 spines at base of coxite, set on distinctly raised tubercles on a slight lobe; another slender spine on inner margin of coxite near or beyond middle.

LARVA
  Seta 1 of antennae always branched. Head setae 2 set close together.

Anopheles (Anopheles) stigmaticus Skuse

Anopheles stigmaticus Skuse, 1889, *Proc. Linn. Soc. N.S.W.*, 3: 1758.

A dark brown mosquito with dark scales on the wings and legs, except on the basal three-quarters of hind femur, which is pale scaled. The knob of the haltere is black scaled.

ADULT FEMALE
  Vertex with light-golden upright scales. Palps almost as long as proboscis; both dark brown. Scutum with light-golden bristles, becoming darker above wing root. Pleura devoid of scales; two or more upper sternopleural bristles. Wing brown scaled, with conspicuous darkening of wing membrane near tip of Sc. Knob of haltere black scaled. Legs dark brown; fore and mid femora pale beneath; basal three-quarters of hind femora

entirely pale, remainder dark (fig. 15d). Abdomen dark brown except segments VII and VIII which are pale.

ADULT MALE

Palps as long as proboscis, two apical segments swollen. *Terminatia* (fig. 15 a-c). Coxite short, blunt tapering, about twice as long as broad, with long setae but devoid of scales; strong subapical spine on inside; stout parabasal spine on elongate base. Style slightly longer than coxite, widened at both ends; terminal appendage small. Ventral lobe of harpago with 4 setae. Phallosome with about 15-17 pairs of slender, smooth leaflets, the longest curved.

LARVA (fig. 15 e-h)

Pale with black dorsal longitudinal stripe on thorax and on abdominal segments I-VI. Head dark brown. Antennal seta 1 short, 3-4 branched; seta 4, single. Pro-, meso- and metathoracic setae 9, single. Metathoracic seta 1 plumose with about 12 slender leaflets. Segments I-VII of abdomen each with narrow transverse tergal plate; behind this, on segments I-VII or III-VIII, a pair of small plates. Pecten with 17-21 spines. Median plate of scoop broadened posteriorly; seta 13, single, very short. Anal segment: seta 1, single, simple; 2 and 3 single, plumose; 4, of 15 plumose setae. Anal papillae about three-quarters length of saddle.

BIOLOGY

This species breeds in cold (55-57°F) clean water shaded by trees, and is largely confined to mountainous areas. The larvae have been collected in streamfed pools, at the edges of slowly-running creeks, in pits in drying-out creeks, and in rock pools.

It does not attack man but has been reported feeding on marsupials (Lee *et al.*, 1954).

DISTRIBUTION

In Victoria *An. stigmaticus* has been recorded from: Club Terrace, Weeragua, Tarra Valley, Woods Point, Maroondah, Warburton, Lorne, Carpendeit, Kennedy's Creek, Grampians. It has also been recorded from New South Wales and Tasmania.

Anopheles (Anopheles) pseudostigmaticus Dobrotworsky
Anopheles pseudostigmaticus Dobrotworsky, 1957, *Proc. Linn. Soc. N.S,W.*, 82: 184.

Anopheles pseudostigmaticus is very similar and closely related to *An. stigmaticus* but differs from it as follows:

ADULT FEMALE

Only one upper sternopleural bristle. Knob of haltere pale scaled. Abdominal segment VIII, but not VII, paler than anterior segments.

ADULT MALE

Terminalia (fig. 16 a-c).   Ventral lobe of harpago with 3 setae. Phallosome with about 12-15 pairs of straight leaflets.

LARVA (fig. 16 e-h)

Uniformly brown without distinct dorsal black stripe. Head light brown

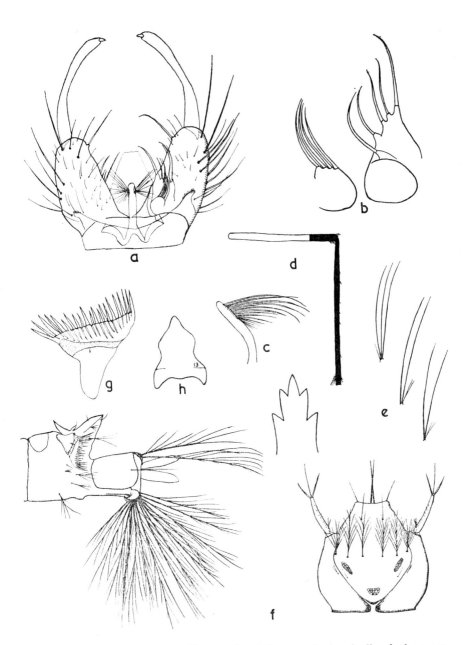

FIG. 15 *Anopheles stigmaticus* Skuse. *a-d*, adult: *a*, male terminalia; *b*, harpago; *c*, phallosome; *d*, hind femur and tibia. *e-h*, larva; *e*, pro- meso- and metathoracic pleural setae; *f*, head, mentum and terminal segments; *g*, pecten; *h*, median plate of scoop. (*c*, after Dobrotworsky.)

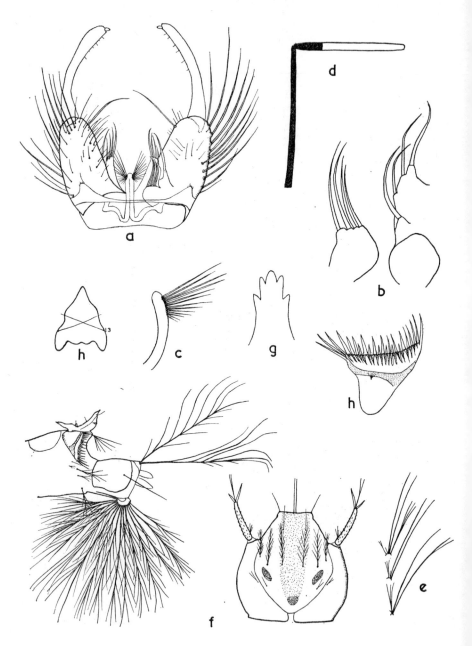

Fig. 16 *Anopheles pseudostigmaticus* Dobrotworsky. *a-d*, adult: *a*, male terminalia; *b*, harpago; *c*, phallosome; *d*, hind femur and tibia. *e-h*, larva: *e*, pro- meso- and metathoracic pleural setae; *f*, head, mentum and terminal segments; *g*, pecten; *h*, median plate of scoop.

with dark spots and median longitudinal stripe. Antennal seta 4, 2 branched. Prothoracic and metathoracic seta 9, 2-3 branched. Pecten with 23-27 spines. Seta 13 of median plate of scoop single, as long as width of plate at origin of seta.

BIOLOGY

Larvae have been collected in the grassy edges of swamps with clean water, in ground pools and rock pools, which usually are exposed to the sun. The water temperature during summer rises to 68-78°F.

DISTRIBUTION

In Victoria *An. pseudostigmaticus* has been recorded from Weeragua, Maroondah and Baxter. It also occurs in New South Wales and Queensland.

Anopheles (Anopheles) atratipes Skuse
Anopheles atratipes Skuse, 1889, *Proc. Linn. Soc. N.S.W.,* 3: 1755.

A dark brown mosquito with dark legs; the wings are dark scaled with silvery sections on some veins and white section of fringe scales at the tip of the wing.

ADULT FEMALE

Integument dark brown with grey bloom. Vertex with pale upright scales, becoming black laterally and towards neck. Palps as long as proboscis, black; scales of palps outstanding at base. Scutum with black setae and scattered narrow pale scales. One upper sternopleural bristle. Wing scales dark brown, C, Sc and R_1 uniformly clothed with dark scales, other veins with silvery scales on some sections; fringe dark, with white section at tip of wing. Haltere with dark knob. Legs dark brown, femora and tibiae paler beneath. Abdomen dark brown, with hairs but devoid of scales.

ADULT MALE

Palps with two terminal segments swollen. *Terminalia* (fig. 17 a, c). Coxite cylindrical, with scales and long setae. Inner surface of coxite with a long subapical seta and with a strong basal tubercle with expanded process. Harpago not distinctly divided into dorsal and ventral lobes, with 9 strong curved spines. Phallosome stout, short, with 3 pairs of terminal leaflets.

LARVA (fig. 17 d-e)

Head very dark, almost black. Antennae swollen on basal half; seta 1 arising half-way along antennae, branched. Head setae: 2, close together, simple; 3, single, simple; 4, small, single or 4 branched; 5, 6 and 7, approximately equal in length, plumose; 8, 3-4 branched; 9, 2-4 branched. Prothoracic setae: 1 and 3 single, simple; 2, plumose. All three pleural groups of three simple setae and one branched apically. Metathoracic palmate (seta 1) reduced to branched hair. Pecten consists of short and long spines. Anal segment: seta 1, single, simple; 2 and 3, branched; 4, of 16 branched setae. Anal papillae equal, about half length of saddle.

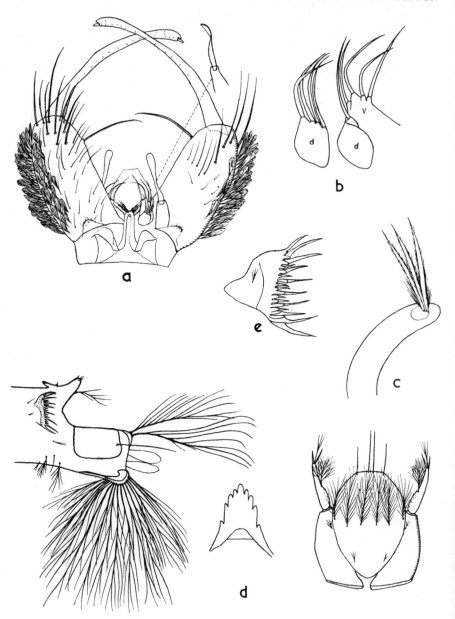

Fig. 17 *Anopheles atratipes* Skuse. *a-c*, adult: *a*, male terminalia; *b*, harpago; *c*, phallosome. *d-e*, larva: *d*, head, mentum and terminal segments; *e*, pecten.

BIOLOGY

Larvae have been found in sluggish, slightly muddy creeks and tea-tree swamps. Adults are present from October till February; they are day-biting and attack man.

DISTRIBUTION

Throughout its range *An. atratipes* is exclusively a coastal species but it is occasionally found inland, e.g. in the Grampians. Victorian records show it occurs from: Cann River in East Gippsland, Cabbage Tree Creek, Wilson's Promontory, Warrnambool, Drik Drik, Milton, Gorac West to Tyrendarra in the Western District. It has also been recorded from Queensland, New South Wales, South Australia and Western Australia.

Subgenus CELLIA Theobald

Cellia Theobald, 1902, *J. Trop. Med.,* 5: 183 and *Mon. Cul.,* 3: 107.
For synonyms see Stone, Knight and Starke (1959).

Characters of the Subgenus

ADULT

Propleural and spiracular hairs variable in number, sometimes absent. Male terminalia: 4-6 parabasal spines, not set on tubercles or lobes; no additional spine on inner margin of coxite near middle. Wings almost always with distinct pale markings.

LARVA

Seta 1 of antennae always short and simple, and usually on outer surface. Head setae 2 separated by distance equal to at least a quarter of their length.

Anopheles (Cellia) annulipes Walker

Anopheles annulipes Walker, 1856, *Ins. Saund. Dipt.,* 1: 433 (A). *Anopheles mastersi* Skuse, 1889, *Proc. Linn. Soc. N.S.W.,* 3: 1757. *Anopheles musivus* Skuse, 1889, ibid., 3: 1754. *Anopheles perplexus* Taylor, 1943, ibid., 68: 153. *Anopheles derricki* Taylor, 1943, ibid., 68: 155, *Anopheles persimilis* Taylor, 1943, ibid., 68: 155.

This is a grey mosquito, the wing and legs profusely marked with white scales. The profuse mottling of white scales on the legs and wings distinguishes this species from all other Victorian Anophelines.

ADULT FEMALE

Vertex with upright forked scales white anteriorly, black laterally and posteriorly. Palps black scaled with broad white bands and white tip (fig. 18 c, d). Proboscis entirely black or with pale scaling on apical half. Integument of thorax dark with greyish bloom and clothed with white scales and dark hairs. Posterior pronotum smooth and bare. Small patches of white scales on lower and upper parts of sternopleuron and near middle of mesepimeron. Wings profusely spotted with black and white patches of scales. Legs (fig. 18 e) with many bands and spots of black and white scales. Fore and mid tarsi with segments 4 and 5 completely black; all segments of hind tarsi black or banded. Abdomen hairy, some scales usually on segments VII and VIII and sometimes on other segments.

ADULT MALE

Palps with patches of white scales on all segments except first. *Terminalia* (fig. 18 a, b). Coxite short with long setae and scales laterally, four strong hooked parabasal spines; accessory spine above them; parabasal lobe absent. Harpago with usual club, one strong seta longer than club, and several shorter setae. Phallosome tubular, tapering towards apex, which bears 4-5 leaflets on each side.

LARVA (fig. 18 f, g)

Dark with variable pattern on dorsal side. Setae very variable. Antennae shorter than length of head; seta 1, small, single, arising near mid length of antenna; seta 4, usually 2 branched. Head setae: 3, branched; 4, small, 3-6 branched; 7, about half length of 5; 8 and 9, 2-15 branched. Prothoracic setae: 1, 2, 4, 5 and 7, plumose; 3 and 6, single, simple. Pecten very variable; it may have a few long spines ventrally and short ones elsewhere or mixed long and short spines along the whole length; spines with minute secondary spines at base. Median plate of scoop broad anteriorly; no lateral projections present. Anal papillae equal; length varies from half length of saddle to 1½ times its length.

BIOLOGY

An. annulipes breeds in a variety of pools and swamps. The larvae have been found in ground and rock pools usually with some vegetation and either shaded for part of the day or fully exposed to the sun. The water may be fresh, brackish, clear or slightly muddy. In elevated forest country *An. annulipes* usually breeds in pools exposed to the sun and in such localities it overwinters in the larval stage. At lower altitudes it breeds throughout the year.

It bites after sunset, attacking man and a wide variety of animals (cow, dog, rabbit, horse and fowl).

DISTRIBUTION

An. annulipes occurs throughout Victoria and is extremely abundant along the Murray River. It is widely distributed in Australia including Tasmania, and is the commonest *Anopheles*. It has also been recorded from New Guinea.

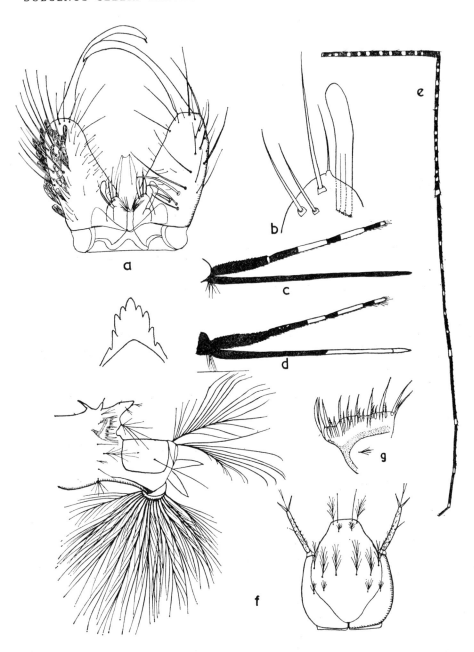

FIG. 18 *Anopheles annulipes* Walker. *a-e*, adult: *a*, male terminalia; *b*, harpago; *c* and *d*, variations in ornamentation of females palps and proboscis; *e*, hind leg. *f-g*, larva: *f*, head, mentum and terminal segments; *g*, pecten.

SUBFAMILY CULICINAE

This subfamily is represented in Victoria by six genera: *Tripteroides, Mansonia, Aedeomyia, Aedes, Culiseta* and *Culex.*

Characters of the Subfamily

ADULT

Clypeus longer than broad, rounded anteriorly. Proboscis slender, not hooked. Palps in females shorter than proboscis. Scutellum trilobed, each lobe bearing bristles. Abdomen densely covered with broad scales (except in male of *Aedes nivalis*). Male terminalia: paraproct always well developed; phallosome never with leaflets at its tip.

LARVA

Head rarely rotatable. None of hairs of body are pinnately branched as in Anophelinae. Abdomen without float hairs; VIIIth segment with siphon and lateral comb.

EGGS

Variously shaped, but never rounded-oval in shape nor provided with lateral floats, as in Anophelinae.

KEYS TO VICTORIAN GENERA OF THE SUBFAMILY CULICINAE

Adults

1   Pulvilli present .. .. *Culex*
      Pulvilli absent 2
2(1) Postspiracular bristles (at least one or two) present; claws of female usually toothed 3
      Postspiracular bristles absent; claws of female simple 4
3(2) Wing scales all broad; female claws simple; base of subcostal vein bare beneath; tip of abdomen blunt (in part) .. .. .. *Mansonia*
      Wing scales usually narrow; if broad, claws of female toothed; base of subcostal vein scaled .. .. *Aedes*
4(2) Spiracular bristles present 5
      Spiracular bristles absent 6
5(4) Several upper sternopleural bristles; base of subcostal vein hairy beneath; tergites unbanded and without lateral patches .. *Culiseta*
      One or two upper sternopleural bristles; base of subcostal vein bare beneath; tergites with lateral white patches .. .. .. *Tripteroides*
6(4) All segments of female antenna and last two of male antenna short and thick; mid and hind femora with scale tuft .. ... *Aedeomyia*

Segments of female and male antennae normal, slender; femora without scale-tuft (in part) .... *Mansonia*

Larvae (fourth stage)

1   Distal half of siphon modified for piercing roots of aquatic plants .... *Mansonia*
      Siphon not modified, normal in shape 2
2(1) Anal segment with ventral brush (seta 4) reduced to one pair of setae .... *Tripteroides*
      Anal segment with ventral brush (4) of several tufts 3
3(2) VIIIth abdominal segment with lateral chitinous plate bearing a row of 19-21 comb teeth on margin .... *Aedeomyia*
      VIIIth abdominal segment usually without lateral plate; if present, only about 5 comb teeth 4
4(3) Siphon with pair of basal tufts (in part) .... *Culiseta*
      Siphon without basal tufts 5
5(4) Siphon with several pairs or median row of ventral tufts .... *Culex*
      Siphon with pair of median or subapical tufts or single setae 6
6(5) All setae non plumose; anal segment completely ringed by saddle (in part) .... *Culiseta*
      Some of setae plumose, anal segment usually not completely ringed by saddle .... *Aedes*

TRIBE SABETHINI

Genus Tripteroides Giles

Tripteroides Giles, 1904, *J. trop. Med.*, 7: 369. *Colonemyia* Leicester 1908, *Cul. Malaya*: 233.

The genus is distributed from India and Ceylon through the Oriental and Australian Regions eastwards into the Pacific to Fiji and New Zealand and northwards to Japan. In Victoria it is represented by two species and one form of uncertain status.

Characters of the Genus

ADULT

Vertex with broad flat scales; a few upright ones on nape. Male palps variable in length. Scutal scales variable in shape, scutellar scales always broad and flat. Usually one posterior pronotal bristle. A few spiracular bristles; not more than two on upper sternopleuron and only a few on the upper mesepimeron; no lower mesepimeral bristles. Pleura usually largely covered with broad flat scales. Wings: fringe of squama may be reduced. Legs slender; first hind tarsal segment usually longer than tibia. Front claws of male unequal, of female equal and simple. Hind tarsi sometimes with only one claw. No pulvilli. Abdomen with a few hairs; VIIIth segment of females broad and very bristly. *Male terminalia.* Coxite with small basal lobe bearing bristles. Style long, slender with short apical appendage. Paraproct with vertical row of strong teeth at tip; phallosome a simple, incomplete tube. IXth tergite with prominent lobes bearing strong bristles.

LARVA

Head small, antennae short, seta 1 small and arising beyond middle. Thorax and abdomen usually bearing stellate setae. Mesothorax usually with long strong spine on a plate. *Abdomen*. VIIIth segment: lateral comb of one row of teeth; seta 1 tuft much more strongly developed than other setae. Siphon with numerous scattered hairs and spines; the latter may form a 'false pecten'. Anal segment: seta 4 represented by a single pair.

Belkin (1962) divided the genus *Triperoides* into three subgenera, only one of which, *Rachionotomyia* is represented in Victoria.

Subgenus RACHIONOTOMYIA Theobald

Rachionotomyia Theobald 1905, *J. Bombay nat. Hist. Soc.*, 16: 248. *Polylepidomyia* Theobald 1905, *Ann. hist. nat. Mus. hung.*, 3: 118. *Skeiromyia* Leicester 1908, *Cul. Malaya*: 248. *Squamomyia* Theobald 1910, *Rec. Indian Mus.*, 4: 28. *Mimeteomyia* Theobald 1910, *Mon. Cul.*, 5: 210. *Tricholeptomyia* Dyar and Shannon 1925, *Insec. Inscit. Menst.*, 13: 72.

Characters of the Subgenus

ADULT

Without silvery markings on thorax and legs. Palps of male almost as long as proboscis. Usually one posterior pronotal bristle. Outstanding scales of all veins narrow, usually rather long and fairly dense.

LARVA

Maxillae without strong apical horns. Dorsolateral plate of metathorax with or without conspicuous spike. Comb teeth in one row.

ATRIPES GROUP

Characters of the Group

Proboscis shorter than abdomen. Male palps almost as long as proboscis. Mesothorax and metathorax of larva with modified spines, lateral comb arising from sclerotized plate.

Tripteroides (Rachionotomyia) atripes (Skuse)

Culex atripes Skuse, 1889, *Proc. Lin.. Soc. N.S.W.*, 3: 1750. *Mimeteomyia apicotriangulata* Theobald, 1910, *Mon. Cul.*, 5: 211. *Rachionotomyia atripes* Edwards, 1924, *Bull. ent. Res.*, 14: 362. *Tripteroides atripes* Lee, 1945, *Proc. Linn. Soc. N.S.W.*, 70: 261.

A small black mosquito without a white pattern on the scutum; the pleura are densely clothed with white scales; the abdomen is unbanded and the tarsi are black.

Adult Female

Vertex black scaled with narrow silvery border to eyes. Proboscis and palps black, the former shorter than abdomen. Scutum clothed with narrow light-bronze scales; patch of elongate white scales in front of wing root. Scutellum with broad light-coloured scales. Usually two spiracular bristles. Posterior pronotum with broad white scales below, bronzy-grey above. Pleura clothed with flat white scales. Legs dark scaled except femora and tibiae, which are pale scaled ventrally. Tergites (fig. 19 c) with large apical lateral triangular patches of white scales visible from above. Sternites white scaled.

Adult Male

Proboscis pale scaled ventrally on basal half or third. Palps about four-fifths length of proboscis; some pale scales ventrally or dorsally. *Terminalia* (fig. 19 a, b). Style as long as coxite. Lobes of IXth tergite not widely separated; outer margin longer than inner and with about 5 long strong setae.

Larva (fig. 19 d-f)

Antennal seta 1, stout, 2 branched. Head setae: 4, 1-2 branched; 5, 2-4 branched; 6, 2-5 branched; 7, 2-5 branched; 8, single; 9, 2 branched. Stellate tufts of thorax and abdomen strongly developed. Mesothoracic and metathoracic spines present. *Abdomen*. VIIIth segment: lateral comb of 5-7 strong blunt-tipped spines, arising from sclerotized plate, continued to ventral surface as row of finer spines barbed on apical third; seta 1, strong stellate tuft with about 20 branches; 2 and 4, single; 3, usually 4 branched; 5, 4-6 branched. Siphon with 11-13 ventral tufts; index 2·2-2·7; dorsal setae mostly 2 branched; pecten of 4-5 spines on basal half of siphon. Anal segment: saddle with strong fringe of spines; seta 1, 3-5 branched; 2, 5-7 branched; 3, single; 4, 5-8 branched. Anal papillae short, pointed.

Biology

The larvae are found in tree holes and water tanks. This as a day-biting mosquito that attacks man.

Distribution

T. atripes occurs in the northern part of Victoria. It has also been recorded from Queensland, New South Wales and South Australia.

Southern Form

On and south of the Central Highlands in Victoria there is found a *Tripteroides* which, though similar in many respects to *T. atripes* from Northern Victoria, exhibits some quite constant morphological differences. The taxonomic status of this southern form cannot be determined until more material is available for study.

The main distinctive morphological features of this form are:

Adult

Proboscis in male black ventrally or with only a few pale scales at base. Scutal scales dark bronze. Lateral patches on tergites (fig. 20 c) small and visible from above only on segments VI and VII.

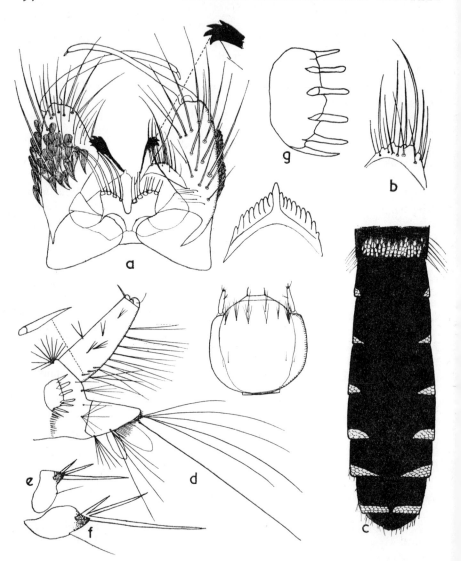

FIG. 19 *Tripteroides atripes* (Skuse). Mildura form. *a-c*, adult: *a*, male terminalia; *b*, basal lobe of coxite; *c*, female abdomen, dorsal view. *d-g*, larva: *d*, head, mentum and terminal segments; *e*, mesothoracic and *f*, metathoracic spines; *g*, lateral comb.

LARVA (fig. 20 d-g)

Lateral comb with 6-9 spines arising from plate; they are more pointed than in the typical form. Pecten of 5-9 spines along siphon, the most apical near its tip.

DISTRIBUTION

In Victoria it has been collected at the following localities: Grampians, Maryborough, Christmas Hills, Wattle Glen, Maroondah and Lorne.

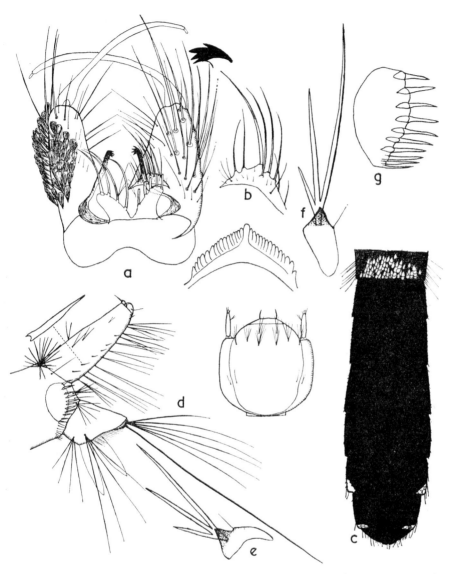

FIG. 20 *Tripteroides atripes* (Skuse). Southern form. *a-c*, adult: *a*, male terminalia; *b*, basal lobe of coxite; *c*, female abdomen, dorsal view. *d-g*, larva: *d*, head, mentum and terminal segments; *e*, mesothoracic and *f*, metathoracic spines; *g*, lateral comb.

CALEDONICUS GROUP

Characters of the Group

Proboscis longer than abdomen. Pleural scales reduced, a conspicuous band of white scales from anterior pronotum across lower part of posterior pronotum to upper part of mesepimeron. Mesothorax and metathorax of larva without modified spines; VIIIth segment: comb without plate.

Tripteroides (Rachionotomyia) tasmaniensis (Strickland)

Stegomyia tasmaniensis Strickland, 1911, *Entom.*, 44: 249. *Rachionotomyia tasmaniensis* Edwards, 1924, *Bull. ent. Res.*, 14: 362. *Rachionotomyia cephasi* Edwards, 1923, *Bull. ent. Res.*, 14:8. *Tripteroides tasmaniensis* Edwards, 1932, *Dipt. Fam. Cul. Gen. Insect.*, 194: 77.

This species is readily distinguished from *T. atripes* by having a band of white scales across the pleuron, and by the pale scaling of the terminal segments of the hind legs.

Adult Female

Vertex black scaled with narrow silvery border to the eyes. Proboscis and palps violet-black. Scutum clothed with narrow brown scales; patch of elongate white scales in front of wing root. Posterior pronotum with white scales below, dark above. 2-4 spiracular bristles. Pleura (fig. 21 c) with band of white scales from anterior pronotum across lower part of posterior pronotum to upper part of mesepimeron; two patches of white scales on sternopleuron, one at middle, one lower, towards posterior border; one patch on lower part of mesepimeron. Legs (fig. 21 d) black with white knee spots and tibial tips; last 2-3 tarsal segments pale scaled. Tergites black scaled with small lateral patches of white scales. Sternites white scaled with some admixture of black scales.

Adult Male

Palps about five-sixths length of proboscis. Terminalia (fig. 21 a, b). Coxite rather long and narrow; basal lobe with about 12 long strong setae; style narrow, about as long as coxite. Paraproct with 4-5 teeth at tip. Lobes of IXth tergite with 8-10 strong, long setae.

Larva (fig. 21 e, f)

Antenna short; seta 1 single, fine and small. Head setae: 4, 2-3 branched; 5, 2-4 branched; 6, single or 2 branched; 7, 3-6 branched; 8, single; 9, single or 2 branched. No spines on thorax. *Abdomen.* VIIIth segment: lateral comb a single row of 17-22 blunt fringed spines; seta 1, stellate tuft; 2 and 4, single; 3, 3-6 branched; 5, usually 3 branched. Siphon with index 2.9-3.1; pecten of 5-8 small spines; 2 basal setae 4-5 branched, 10-13 ventral setae 2 branched; dorsal setae single or 2 branched. Anal segment: seta 1, single or 2 branched: 2, 5-7 branched; 3, single; 4, 4-5 branched. Anal papillae more than three times as long as saddle, with rounded tips.

Biology

T. tasmaniensis breeds in tree holes, discarded tyres, tins and sometimes in rock pools. It overwinters in the larval stage. This is a day-biting mosquito that attacks man.

Distribution

T. tasmaniensis is common in Victoria in the elevated woodlands with an average annual rainfall of not less than 30 inches. It also occurs in New South Wales and Tasmania.

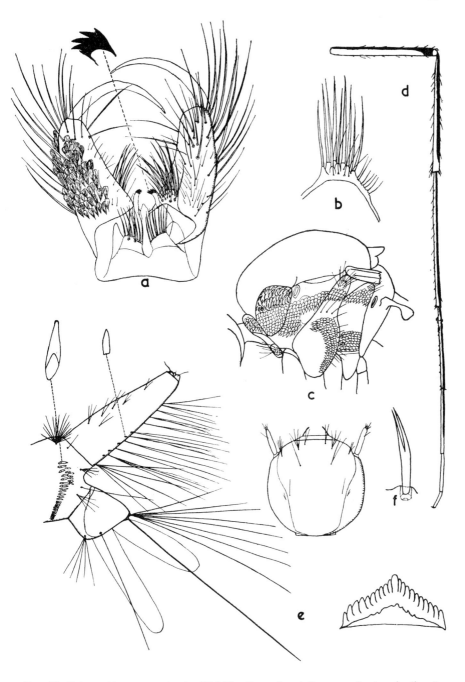

Fig. 21 *Tripteroides tasmaniensis* (Strickland). *a-d*, adult: *a*, male terminalia; *b*, basal lobe; *c*, side view of thorax; *d*, hind leg. *e-f*, larva: *e*, head, mentum and terminal segments; *f*, head seta 1.

ARGENTEIVENTRIS GROUP

Characters of the Group

Proboscis longer than abdomen. Pleural scaling extensive. Male palps about three-quarters length of proboscis.

Tripteroides (Rachionotomyia) marksae Dobrotworsky n. sp.

Types

The type series was bred from larvae collected (22.1.63) in a jar placed in a tree stump at Cabbage Tree Creek, Victoria. The holotype, allotype, seven paratype males and six paratype females are in the collections of the National Museum, Melbourne. One paratype male and one paratype female are in each of the following collections: Australian National Insect Collection, Canberra; School of Public Health and Tropical Medicine, Sydney; University of Queensland, Brisbane; British Museum (Natural History), London; U.S. National Museum, Washington. All have associated larvae and pupal skins.

This species is named in honor of Dr E. N. Marks in recognition of her contributions to the knowledge of the mosquitoes of the Australian Region.

This species is similar to *T. atripes* but has a lighter integument and the proboscis is longer than the abdomen.

Adult Female

Vertex light golden-brown scaled. Proboscis and palps black; proboscis longer than abdomen. Palps about one-twelfth length of proboscis; Scutum clothed with narrow dark-bronze scales; elongate patch of larger whitish scales in front of wing roots. Scutellum with narrow dark scales. Usually two spiracular bristles. Posterior pronotum with elongate white scales. Pleura extensively clothed with broad white scales. Legs: coxae and trochanters pale. Femora, tibiae and tarsi dark scaled anteriorly, light scaled posteriorly; light scaling on fore femur extending anteriorly near the middle, leaving there only a black line. Tergites dark scaled with white lateral patches visible from above only on segments VI and VII. Venter white scaled, except sternite VIII, which is black scaled.

Adult Male

Vertex scaled with darker scales than in female. Palps about three-quarters length of proboscis. *Terminalia* (fig. 22 a, b). Coxites with numerous black scales and bristles. Style curved, almost as long as coxite. Paraproct with five teeth. Lobes of IXth tergite with 9-10 long, strong setae.

Larva (fig. 22 c-e)

Antenna short; seta 1, 10-15 branched. Head setae: 4, 9-12 branched; 5, 6 and 7, 13-18 branched, all plumose; 8, single; 9, 2 branched. Mentum with long central tooth and 8 or 9 lateral teeth. Stellate tufts of thorax and abdomen strongly developed. Metathoracic spines with one long and three short branches. *Abdomen:* VIIIth segment: lateral comb a single row of 30-35 spines; seta 1, a strong stellate tuft; 2 and 4 single; 3, 5-8 branched; 5, 5-6 branched. Siphon spinose, with index $3 \cdot 8$-$4 \cdot 1$; pecten

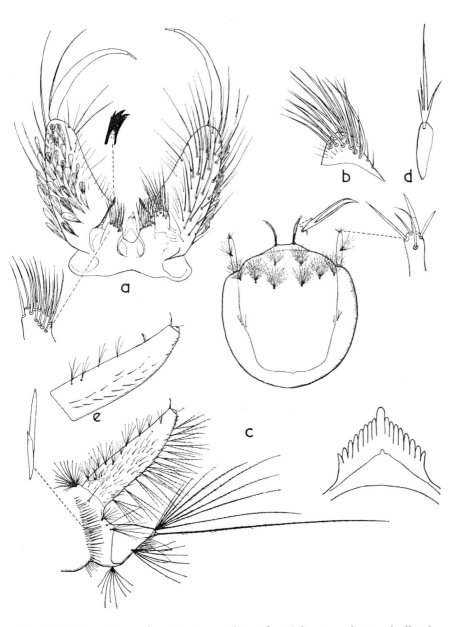

Fig. 22 *Tripteroides marksae* Dobrotworsky. *a-b*, adult: *a*, male terminalia; b, basal lobe. *c-e*, larva: *c*, head, mentum and terminal segments; *d*, metathoracic spine; *e*, siphon.

of 8-13 spines; basal setae 4-5 branched; 12-13 ventral setae, 3 branched; dorsal setae 2-10 branched. The number of pecten-like spines on siphon is very variable; in some larvae they are numerous on basal three-quarters of siphon, in others there are only a few (fig. 22 c, e). The number and the branching of the dorsal setae also vary a great deal. Anal segment: seta 1, 5-6 branched; 2, 7-8 branched; 3, single; 4, 7-8 branched. Anal papillae about as long as saddle. Saddle with a strong fringe of long fine spines distally.

BIOLOGY AND DISTRIBUTION

Larvae of this species have been found in a jar containing water and decayed wood placed in a hollow tree stump. It is known only from the type locality: Cabbage Tree Creek, Victoria.

TRIBE CULICINI

Genus MANSONIA Blanchard

Taeniorhynchus Lynch Arribalzaga, 1891, *Rev. Mus. La Plata*, 1: 374; 2: 147 (Generic name suppressed, 1959, Int. Comm. Zool. Nomencl.). *Panoplites* Theobald, 1900, *Rept. Coll. Brit. Mus.*, 5 (*non* Gould 1853). *Mansonia* Blanchard, 1901, *C.R. Soc. Biol.*, 53: 1045. *Coquillettidia* Dyar, 1905, *Proc. ent. Soc. Wash.*, 7: 45. *Mansonioides* Theobald, 1907, *Mon. Cul.*, 4: 498. *Rhynchotaenia* Brethes, 1911, *Ann. Mus. Nac. Buenos Aires*, 13: 470. *Pseudotaeniorhynchus* Theobald, 1911, *Novae Culicidae*, p. 19.

The genus, which is world-wide in distribution, is represented in Victoria by four species: *M. linealis* is widely distributed but the other three are restricted, *M. uniformis* to the northern part, *M. aurata* and *M. variegata* to East Gippsland.

Characters of the Genus

ADULT

Proboscis not swollen at tip. Palps of male as long as proboscis or longer, of female not more than a quarter as long as proboscis. Vertex with numerous upright forked scales and narrow curved decumbent scales. Postspiracular bristles present or absent; lower mesepimeral bristles usually present. Male terminalia variable in structure; harpago short or long, bearing one or two spines at tip. VIIIth segment of female always short and broad; cerci short. Claws of fore and mid legs of male unequal, larger claw usually with teeth, smaller usually simple. All claws of female simple. No pulvilli. Wings may have broad scales.

LARVA

Antennae long, with large branched seta 1 arising before middle; setae 2 and 3 inserted well before tip. Mentum small. Pair of large tracheal dilatations in thorax. Lateral comb of a few long teeth in one row. Siphon short, without pecten, with one pair of setae. Valves modified into piercing apparatus. Saddle complete ring.

The adults provide good morphological traits for division of the genus into four subgenera; larvae show few variations and are very similar in all species.

Two subgenera *Coquillettidia* and *Mansonioides* are represented in Victoria.

Subgenus COQUILLETTIDIA Dyar

Coquillettidia Dyar, 1905. *Proc. ent. Soc. Wash.*, 7: 45.

Characters of the Subgenus

ADULT

Postspiracular bristles absent. Male palps with last segment large and penultimate segment not upturned. Male terminalia: style modified, but always with short terminal appendage; harpago with a spine as long as, or longer than, harpago itself; paraproct with several teeth; phallasome divided into lateral plates. Abdomen: segment VII large; segment VIII small, without sclerotized hooks. Wing scales rather narrow. Tarsi banded.

LARVA

Distal part of antenna as long as or longer than basal; setae 2 and 3 rather short and placed near middle of antenna; head setae 5 and 6 short tufts.

Mansonia (Coquillettidia) linealis (Skuse)

Culex linealis Skuse, 1899, *Proc. Linn. Soc. N.S.W.*, 3: 1747. *Taeniorhynchus linealis* Edwards, 1924, *Bul. ent. Res.*, 14: 366. *Mansonia linealis* Edwards, 1932, *Dipt. Cul. Genera Insect.*, 194: 118.

The scutum is clothed with dark-bronze scales except light-golden longitudinal lines; the tarsi have inconspicuous basal bands on 2-3 segments; the sternites are white scaled with a median black patch and a black apical border.

ADULT FEMALE

Vertex with narrow curved and upright forked creamy scales, the latter becoming dark laterally. Palps and proboscis black. Scutum (fig. 23 b) clothed with narrow dark-bronze scales with longitudinal lines of light-golden scales; a few broad white scales in front of wing root. Posterior pronotum with narrow light-golden scales. Sternopleuron and mesepimeron with rather large patches of white scales; usually 3 lower mesepimeral bristles. Wing scales dark (fig. 23 c). Fore and mid femora mottled; hind femora (fig. 23 d) white scaled with increasing black mottling towards apex. Knee spots conspicuous. Tibiae dark scaled with white line. Tarsi dark with white line on first segment and inconspicuous basal bands on first 2 or 3 segments of hind legs. Tergites black scaled, usually unbanded with or without narrow patches of white scales basally; lateral patches white. Sternites white scaled with median black patches and black apical border.

Adult Male

Palps as long as or slightly shorter than proboscis. Tergites with narrow creamy basal bands. Fore and mid claws unequal, the larger with two teeth, the smaller with one. *Terminalia* (fig. 23a). Coxite short, almost cylindrical, less than twice as long as broad, with long and short setae, and black scales sternally. Style narrow basally but very broad apically, with small terminal appendage. Harpago with a long rod-like spine and a seta at tip. Paraproct with 4-5 apical teeth. Phallosome with 4-5 teeth on tip of each plate. Lobes of IXth tergite with 2-3 strong setae.

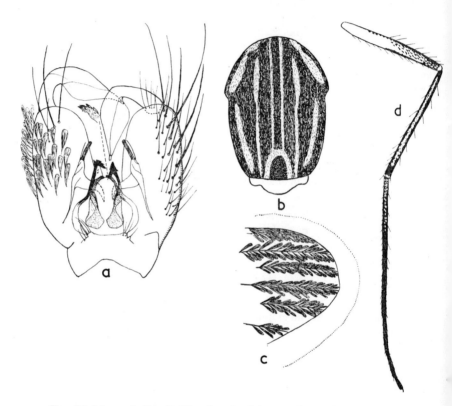

Fig. 23 *Mansonia linealis* (Skuse). *a-d*, adult: *a*, male terminalia; *b*, thorax; *c*, portion of wing; *d*, hind leg.

Larva

Unknown.

Biology

M. linealis is a vicious day-biting mosquito which attacks man as well as domestic animals. Adults are usually common from December to March. In the laboratory, eggs were laid in rafts on the water surface; but it was not possible to rear the larvae; they failed to attach themselves to plant roots and died within 3-4 days, before reaching the second stage.

DISTRIBUTION

In Victoria *M. linealis* is widely distributed. It has also been recorded from Queensland, New South Wales and South Australia.

Mansonia (Coquillettidia) aurata Dobrotworsky
Mansonia aurata Dobrotworsky, 1962, *Proc. Linn. Soc. N.S.W.*, 87: 295.

The scutum is uniformly clothed with light-golden scales; the tarsi have basal creamy bands; the venter is pale with a few dark scales.

ADULT FEMALE

Vertex with narrow curved and upright light-golden scales, the latter becoming dark laterally. Palps and proboscis violet-black scaled with some admixture of pale scales. Scutum clothed with light-golden narrow curved scales and some black bristles. Posterior pronotum with narrow light-golden scales. Postspiracular area with patch of pale scales, large patch of pale scales on sternopleuron and mesepimeron; 3-5 lower mesepimeral bristles. Wing with narrow violet-black scales. Leg (fig. 24 e): fore femur pale with mottling of violet-black scales increasing towards apex; mid femur with a few white scales mostly violet-black anteriorly; hind femur creamy scaled with dorsal black stripe widening towards apex. Tibiae violet-black, those of fore and mid legs with some scattered pale scales. First tarsal segments of all legs with some pale scales on basal half. Fore tarsi with 2 basal white bands, mid with 2-3, hind with 3 or 4 basal white bands. Tergites violet-black with lateral patches of white scales. Sternites white with some admixture of black scales.

ADULT MALE

Palps exceed length of proboscis including labella by two-thirds of terminal segment. *Terminalia* (fig. 24 d). Coxite slightly more than twice as long as broad, scaled sternally and laterally and with long setae apically and laterally. Style with large membranous expansion in middle. Harpago with blunt rod-like spine and seta at apex. Paraproct with 5-6 strong teeth. Phallosome with several small teeth on tip of each plate. Lobes of IXth tergite with 4-5 setae.

LARVA

Unknown.

BIOLOGY

M. aurata is a day-biting species which attacks man. In the laboratory eggs were laid in rafts on the water surface. As with *M. linealis*, the larvae died within a few days.

DISTRIBUTION

M. aurata is known from East Gippsland, Victoria (Cabbage Tree Creek, Cann River and Genoa), where it was collected during December-February.

Mansonia (Coquillettidia?) variegata Dobrotworsky
Mansonia variegata Dobrotworsky, 1962, *Proc. Linn. Soc. N.S.W.*, 87: 293.

This species is readily distinguished from other members of the genus by having mottled wings and the tarsal bands apical.

ADULT FEMALE

Vertex with narrow curved pale scales; upright scales pale medially, black laterally and towards neck. Proboscis creamy scaled, except for black scaled base and tip. Integument dark brown. Pleura with dark band extending from upper part of mesepimeron to posterior pronotum. Scutum

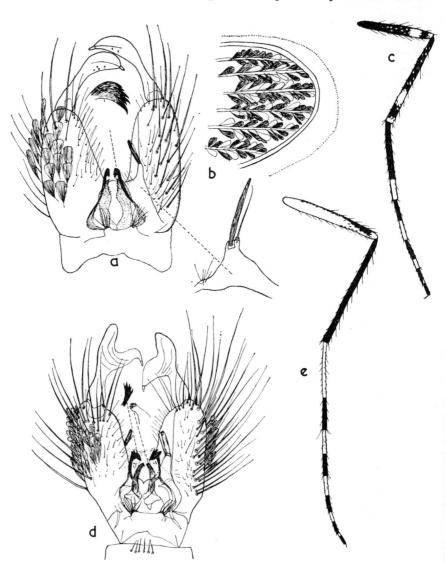

FIG. 24 *Mansonia variegata* Dobrotworsky. *a*, male terminalia; *b*, portion of wing; *c*, hind leg. *Mansonia aurata* Dobrot. *d*, male terminalia; *e*, leg. (*a*, after Dobrotworsky.)

clothed with narrow curved light-golden and black scales. Posterior pronotum with narrow curved light-golden scales. Two patches of broad scales on sternopleura. Two lower mesepimeral bristles. Femora and tibiae mottled and with preapical patch of pale scales, most conspicuous on hind legs. Knee spots white. First three tarsal segments of fore and mid legs with white apical bands. Hind tarsi with four apical bands. Wings (fig. 24 b) with broad scales, mottled. Tergites black scaled with white basal lateral patches; VIIIth tergite with some pale scales. Sternites black with small lateral patches of white scales.

ADULT MALE

Palps exceed length of proboscis, including labella, by terminal segment. *Terminalia* (fig. 24 e). Coxite slightly more than twice as long as broad; sternally and laterally with broad scales, tergally and laterally with long setae. Harpago with long spine accompanied by seta. Style swollen distally, with small appendage. Paraproct with 5-6 strong teeth. Phallosome scoop-shaped with thickened margins and teeth along each side. Lobes of IXth tergite small, with 5 seta each.

LARVA

Unknown.

BIOLOGY

Mating has been observed just after sunset; it is initiated in flight, and completed on grass. *M. variegata* attacks man after sunset.

DISTRIBUTION

M. variegata is known from Cann River, Victoria, and also occurs at Nelson Bay, New South Wales, and Noosa (90 miles north of Brisbane), Queensland.

Subgenus MANSONIOIDES Theobald

Mansonioides Theobald, 1907, *Mon. Cul.*, 4: 498.

Characters of the Subgenus

ADULT

Palps of males only moderately hairy; penultimate segment rather long and strongly upturned, terminal segment minute. Postspiracular bristles present; no postspiracular scales. Wing scales very broad. Male terminalia: harpago in form of long arm with short terminal appendage at its tip; paraproct with several terminal teeth; phallosome simple, undivided. Female abdomen: segment VII, reduced, much smaller than VI; segment VIII, small, concealed within VII, its tergite with terminal row of small hooks or teeth; cerci with upturned pointed tips. Two large spermathecae; third small, pale or absent.

LARVA

Antenna of moderate length; setae 2 and 3 very long and placed beyond middle of antenna; terminal portion less slender than in the other subgenera. Head setae 5 and 6 minute, scarcely distinguishable.

Mansonia (Mansonioides) uniformis (Theobald)

Panoplites uniformis Theobald, 1901, *Mon. Cul.*, 2: 180. *Panoplites australiensis* Giles, 1902, *Handb. Gnats*, 2nd ed.: 355. *Taeniorhynchus uniformis* Edwards, 1924, *Bull. ent. Res.*, 14: 365. *Mansonia uniformis* Edwards, 1932, *Dip. Cul.*, 120.

This species is readily distinguished from all other Australian mosquitoes by having wings mottled with pale broad scales and six white bands on the hind tarsi.

Adult Female

Vertex with narrow pale scales and upright light brown scales. Palps with dark and white scales. Proboscis with light-ochreous scales on basal three-quarters and some darker ones at base; apical quarter dark. Integument light brown. Scutum clothed with golden scales; some silvery scales on shoulders, on fossa, in front of wing roots and round bare area. Posterior pronotum with pale narrow scales. Postspiracular area with 10-15 setae. Sternopleuron and mesepimeron each with a large patch of broad white scales; usually 5 lower mesepimeral bristles. Wings (fig. 25 c) with broad yellowish and dark scales. Femora dark scaled with some admixture of light scales and with large patches of white scales; tibiae dark scaled with patches of white scales. Tarsi with basal bands and with additional band in middle of first segment. Tergites with dark ochreous scales, and white lateral patches. Sternites with ochreous white scales.

Adult Male

Palps longer than proboscis, with two white rings on shaft and one at base of penultimate segment, terminal segment white scaled. *Terminalia* (fig. 25 a, b). Coxite tergally with row of strong, almost spine-like, setae on inner side. Harpago long, reaching almost to end of coxite, with stout, dark terminal appendage at its tip. Style flattened, curved, pointed, hairy distally on external border. Phallosome about as long as broad.

Larva (fig. 25 d)

Head broad. Antenna spiculate; seta 1, arising about one-third length from base, about 20 branched; setae 2 and 3 equal, arising about two-thirds of length from base. Head setae small; setae 4, 5, 8 and 9, 3-4 branched; 6, 6-8 branched; 7, 8-10 branched. Mentum with 5 lateral teeth on each side. *Abdomen*. VIIIth segment: lateral comb of 2 or 3 long, narrow, blunt spines; setae 1, 3 and 5, 2 branched; 2, about 8 branched; 4, about 4 branched. Siphon very short, with single tuft on each side. Anal segment: seta 1, small, about 5 branched; 2, about 10 branched; 3, about 8 branched; 4, of 8 tufts, four of them precratal. Anal papillae about three-quarters length of saddle.

Biology

Adults have been recorded attacking cattle. No larvae have been found in Victoria.

Distribution

In Victoria it is known only from the Mildura area, but this is a widely

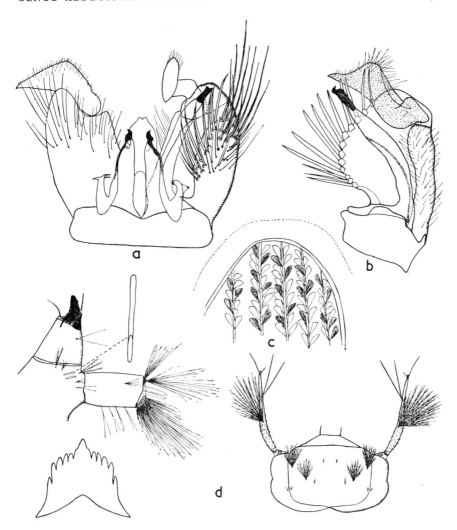

FIG. 25 *Mansonia uniformis* (Theobald). *a-c*, adult: *a*, male terminalia; *b*, coxite, inner lateral aspect; *c*, portion of wing. *d*, larva: head, mentum and terminal segments.

distributed species which has been recorded in the Ethiopian, Oriental and Australian Regions.

Genus AEDEOMYIA Theobald

Aedeomyia Theobald, 1901, *Mon. Cul.*, 1: 98 and 2: 218.
Aedomyia Giles, 1902, *Handb. Gnats*, 2nd ed.: 478 (emend.). *Lepiothauma* Enderlein, 1923, *Wien. ent. Zeit.*, 40: 25.

This genus is mainly tropical in distribution. Of the six known species two are represented in Australia, and one, *Aedeomyia venustipes* Skuse, occurs in Victoria.

Characters of the Genus

ADULT

Palps in both sexes less than a quarter of length of proboscis. Thorax with all scales large and broad. Lower mesepimeral bristles present. Scutellum with a few bristles on middle lobe. Mid and hind femora with apical tufts of sub-erect scales. Fourth tarsal segment of front and mid legs in both sexes shorter than fifth; first hind tarsal segment about as long as tibia. All claws in female equal and simple. No pulvilli. Wing rather short and broad, densely clothed with broad scales. VIIIth segment of abdomen in female short and broad; cerci short. Male terminalia small; coxite short, without distinct lobe; style short and simple, with comb-like terminal appendage; phallosome rounded, without lateral plates; IXth tergite reduced.

LARVA

Integument hairy. Head very large. Antennae large, hairy and broad, with large tuft, three very long apical setae and a stout black spine at tip. Thorax with main hairs extremely long. VIIIth abdominal segment with single row of comb teeth attached to small lateral plate. Siphon short, no pecten; large seta 1 at or beyond middle. Anal segment completely ringed by saddle, hairy; setae 2 and 3 single, strongly plumose along upper edge; seta 4 strongly plumose hairs.

Aedeomyia venustipes Skuse

Aedes venustipes Skuse, 1889, *Proc. Linn. Soc. N.S.W.*,
3: 1761. *Aedeomyia venustipes* Edwards, 1924, *Bull. ent. Res.*, 14: 364. Douglas, 1961, 86: 262.

Is readily distinguished from other Victorian mosquitoes by having two broad white bands on the proboscis, broad scutal scales, and the third tarsal segment of the hind legs entirely white.

ADULT FEMALE

Vertex with broad flat white and black scales; upright scales creamy medially, black laterally. Antennae shorter than proboscis, flagellar segments (fig. 26 c) short, thick and subequal. Palps black, mottled with white scales. Proboscis (fig. 26 b) black scaled with white bands near base, in middle and at tip. Scutum (fig. 26 d) clothed with broad scales: longitudinal ochreous stripe between dorsocentral bristles, widening near middle; laterally and around prescutellar bare area a mixture of black and white scales. Scutellum with broad black and white scales. Posterior pronotum with black and white scales. Pleura with large patches of white scales; 4-6 lower mesepimeral bristles. Wing with large, oval dark scales, mottled with ochreous and white scales; 3 white patches on costa and several on other veins. Femora, tibiae and first tarsal segment with white patches and mottling (fig. 26 e). Fore tarsi with 2 basal bands, mid tarsi with 3; hind tarsal segment 2, with white basal band; segment 3, entirely white; 4, white with narrow apical black band; 5, black. Claws simple. Tergites

GENUS AEDEOMYIA THEOBALD

dark scaled and mottled with white and creamy scales particularly laterally. Sternites black scaled, mottled with white.

ADULT MALE

Palps short, about one-sixth length of proboscis. Venter less mottled

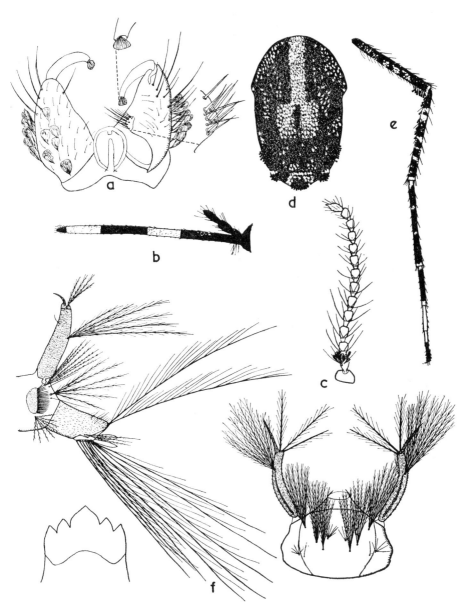

FIG. 26 *Aedeomyia venustipes* Skuse. *a-e*, adult: *a*, male terminalia; *b*, proboscis and palp of male; *c*, female antenna; *d*, thorax; *e*, hind leg. *f*, larva: head, mentum and terminal segments.

than in female. Fore and mid claws large and slightly unequal, larger claw with tooth, smaller simple. *Terminalia* (fig. 26 a). Coxite short, tapering, about twice as long as broad, clothed sternally and laterally with moderate number of long and short setae; basal and apical lobes indistinct; basal patch of strong spine-like setae on inner side of coxite. Style short, stout with comb-like terminal appendage. Phallosome moderately sclerotized, rounded, with lateral plates. IXth tergite reduced.

LARVA (fig. 26 f)

Antennae swollen, curved and hairy; seta 1, long, 8-12 branched. Head setae: 4, 7-12 branched; 5, 5-7 branched; 6, 7-10 branched; 7, 9-10 branched; 8, 3 branched; 9, 5-6 branched. Prothoracic setae: 1, 5, 6 and 7, single; 2, 3 branched; 3, about 15 branched; 4, 2 branched. *Abdomen*, VIIIth segment: lateral comb of 19-21 spines arising from plate; setae 1, 5-6 branched; 2 and 4, single; 3, 4-6 branched; 5, 9-11 branched. Siphon hairy; index 3·7-4·1; seta 1, long, 5-6 branched; dorsolateral seta single; pecten absent; valve seta 8, 5-6 branched; seta 9, strong hook. Anal segment: saddle complete ring, hairy, with spine-like setae dorsally; seta 1, 3 branched; 2 and 3 single, plumose; 4, of 10 or 11 single, plumose setae. Anal papillae narrow, pointed, about half length of saddle.

BIOLOGY

Larvae of this species have been found throughout the year in permanent swamps and well vegetated dams and pools. Larvae and pupae are usually found amongst surface vegetation but have also been collected in deeper parts of reservoirs. Nothing is known about the feeding habits of this species.

DISTRIBUTION

In Victoria *A. venustipes* has been recorded from several localities in Gippsland (Bairnsdale, Box Bridge, Darriman, Won Wron, Giffard, Woodside, Yarram) and also from Melbourne and Ouyen. It has also been recorded in New South Wales.

Genus AEDES Meigen

Aedes Meigen, 1818, *Syst. Beschr.*, 1: 13. For synonymy see Stone, Knight and Starke, 1959.

This large genus has a world-wide distribution and more than a half of the Victorian species of mosquitoes belong to it. The genus is divided into 23 subgenera, six of which are represented in Victoria.

Characters of the Genus

ADULT

Palps of male rarely longer than proboscis, sometimes quite short; palps of female rarely more than a quarter as long as proboscis. Upright scales on vertex usually numerous. Scutum with variable ornamentation. Mesonotal bristles well developed; several prescutellar hairs present. Pronotal lobes widely separated. 4-6 proepimeral bristles. Spiracular bristles absent; postspiracular bristles always present; prealar, sternopleural and upper mesepimeral bristles rather numerous; lower mesepimeral bristles present or absent. Pleura usually extensively scaly. Upper margin of meron above level of base of hind coxa. Male terminalia variable in structure; tip of paraproct without teeth, spines or hairs. VIIIth segment of female abdomen usually more or less retractile. Cerci often long. Hind tibia with a close-set row of hairs on inner side at tip. First segment of hind tarsus distinctly shorter than tibia. Claws of front and mid legs of male unequal, toothed; tarsal claws of females usually toothed at least on front and mid legs. Pulvilli absent or hairlike. Wing membrane with distinct microtrichia.

LARVA

Very variable. Siphon usually short, rarely with index more than 4; with a single pair of siphonal tufts, never inserted near base. Abdomen without dorsal chitinous plates except on anal segment, which has a dorsal saddle, rarely a complete ring.

KEY TO VICTORIAN SUBGENERA OF THE GENUS AEDES

Adult Females

1	Wings with a tuft of very long scales at base	*Chaetocruiomyia*
	Wings without scale tuft	2
2(1)	Claws toothed at least on front and middle legs	3
	Claws all simple	6
3(2)	Cerci long; eighth abdominal segment narrow and completely retractile	4
	Cerci shorter; eighth segment large and not completely retractile	*Finlaya*
4(3)	Cross-veins clouded; M_{3+4} arising distal to r-m	*Mucidus*
	Cross-veins not clouded; M_{3+4} arising proximal to r-m	5
5(4)	Head with narrow scales dorsally	*Ochlerotatus*
	Head with broad flat scales dorsally	*Pseudoskusea*
6(2)	Tarsi banded	*Macleaya*
	Tarsi unbanded	*Halaedes*

KEY TO VICTORIAN SPECIES OF AEDES LARVAE (FOURTH STAGE)

1	Siphon pilose	*dobrotworskyi*
	Siphon not pilose	2
2(1)	Mouth brushes modified for predacity; head setae 4 to 9 all single	*alternans*
	Mouth brushes not modified; at least head seta 7 with two or more branches	3
3(2)	Head seta 5, single (rarely 2 branched on one side)	4
	Head seta 5 with two or more branches	15
4(3)	Lateral comb teeth in irregular or triangular patch	5
	Lateral comb teeth in a single row	9
5(4)	Siphonal seta 1 longer than half length of siphon	6
	Siphonal seta 1 short, much less than half length of siphon	7
6(5)	Body usually pigmented; mentum with 7-8 lateral teeth on each side; distance between base of distal spine of pecten and base of seta 1 smaller than width of siphon at level of seta 1	*bancroftianus*

Body milky-white; mentum with 9-11 lateral teeth on each side; distance between base of distal pecten spine and base of seta 1 greater than width of siphon at level of seta 1 .. *postspiraculosis*
7(5) Head seta 1 almost straight; seta 1 of VIIIth abdominal segment very small tuft; anal papillae equal and shorter than saddle 8
Head seta 1 stout, curved; seta 1 of VIIIth abdominal segment moderately long; anal papillae unequal, upper pair as long as saddle .. .. *notoscriptus*
8(7) Setae 2 and 4 of VIIIth abdominal segment single; saddle almost complete ring; 2 precratal tufts; anal papillae elongate .. *vigilax*
Setae 2 and 4, 3-4 branched; saddle a small dorsal plate; no precratal tufts; anal papillae very small, globular .. .. *australis*
9(4) Lateral comb teeth arising from plate; siphonal seta 1, single; saddle covering not more than half of anal segment; anal papillae large, broad, blunt ended .. .. *tremulus*
Lateral comb without plate 10
10 (9) Pecten with 2-3 detached spines above seta 1 11
Pecten without detached spines 12
11(10) No precratal tufts; anal papillae twice as long as saddle; siphonal seta 1, single or 2 branched .. .. *vittiger*
Two or three precratal tufts; anal papillae shorter or only slightly longer than saddle; siphonal seta 1 multibranched .. .. *nigrithorax* or *sagax*
12(10) Head seta 4 a tiny tuft; scales of lateral comb with long central spine sometimes fringed at extreme base; saddle not reduced; tufts of ventral brush 5-9 branched; anal papillae narrow, pointed 13
Head seta 4 a large tuft; scales of lateral comb in form of spine fringed on basal two-thirds; saddle reduced; tufts of ventral brush 2 branched; anal papillae broad, tip rounded .. *mallochi*
13(12) Lateral comb of 4-7 spines; seta 1 of VIIIth segment single; 2 and 4, 2-3 branched; pecten consisting of closely-set row of spines; anal papillae unequal and shorter than saddle 14
Lateral comb of 10-14 spines; seta 1 of VIIIth segment with more than 6 branches; 2 and 4, single, anal papillae equal, longer than saddle .. .. *imperfectus*
14(13) Clypeus brown with 3 paler lenticular patches; lateral comb of 4 spines; seta 2 of anal segment 10-15 branched .. ... *stricklandi*
Clypeus light brown with 3 darker lenticular patches; lateral comb of 4-7 spines; seta 2 of anal segment 6-10 branched .. *spilotus*
15 (3) Head setae 5 and 6 moderately long 16
Head setae 5 and 6 long (extending beyond tip of antennae), with 2 unequal branches .. .. *multiplex*
16(15) Head seta 6, single 17
Head seta 6, multibranched 21
17(16) Lateral comb of fringed scales without central long spine .. 18
Lateral comb of fringed scales with central long spine 19

GENUS AEDES MEIGEN 73

18(17) Lateral comb patch of less than 36 scales; siphon index less than 3; pecten of thick spines with pale tips .... *calcariae*
Lateral comb patch of 35-53 scales; siphon index more than 3; pecten spines slender, tips black .... *flavifrons*
19(17) Ventral brush with precratal tufts; anal papillae about as long as, or longer than saddle 20
Ventral brush without precratal tufts; anal papillae short, less then half as long as saddle .... *clelandi*
20(19) Siphon not sclerotized to base dorsally, index more than 3; siphonal seta 1 long; setae 1 and 2 of VIIIth abdominal segment very small .... *theobaldi*
Siphon sclerotized to base dorsally; index 3 or less; setae 1 and 2 of VIIIth segment moderately long; saddle almost complete ring .... *procax*
21(16) Antennae at least as long as head or longer; scales of lateral comb fringed; spines of pecten black with pale tip; anal papillae unequal, shorter than saddle .... *purpuriventris*
Antennae shorter than head 22
22(21) Anal papillae small, globular; scales of lateral comb coarsely fringed, without long central spine .... *camptorhynchus*
Anal papillae elongate, pointed 23
23(22) Lateral comb scales finely fringed 24
Lateral comb scales with 1-4 central longer spines, coarsely or inconspicuously fringed basally 27
24(23) Lateral comb large patch of more than 100 scales, expanded apical part with longer fringe 25
Lateral comb of 19-31 narrow, pointed scales laterally and apically fringed, antennal seta 1, single or 2 branched; seta 1 of anal segment small, fine, 4-5 branched .... *subauridorsum*
25(24) Head setae 4, 5 and 6, almost in straight line; prothoracic seta 4, single; 5, 2-3 branched .... *rupestris* or *subbasalis*
Head setae 4, 5 and 6, forming triangle; prothoracic seta 5, single 26
26(25) Head setae 4, 5 and 6, forming almost a right-angle triangle with seta 4 well in front of seta 5; prothoracic seta 4, single .... *alboannulatus*
Head setae 4, 5 and 6, forming obtuse triangle, with seta 4 slightly in front of or behind seta 5, prothoracic seta 4, 2-3 branched .... *rubrithorax* or *tubbutiensis* or *milsoni*
27(23) Siphon index exceeds 3; pecten of 24-37 close-set, strong, dark spines with paler tip 28
Siphon index usually less than 3; pecten of 16-25 evenly spaced spines with dark tip 29
28(27) Prothoracic setae 1, 2 and 6, single; 3, 4 and 5, 2 branched .... *andersoni*
Prothoracic setae 1, 2, 4, 5 and 6 single; 3, 2 branched .... *continentalis*

29(27) Scales of lateral comb with central tooth at least twice as long as nearest lateral 30
Scales of lateral comb with 2-4 longer spines of almost equal length .. .. *cunabulanus*
30(29) Scales of lateral comb coarsely fringed at base 31
Scales of lateral comb spine-like with inconspicuous fringe at base; siphon index about 3·7 .. .. *perkinsi*
31(30) Anal papillae less than half length of saddle; siphon index 2·9-3·4 .. .. *luteifemur*
Anal papillae almost as long as saddle or longer .. .. *nivalis* or *silvestris*

Subgenus MUCIDUS Theobald

Mucidus Theobald, 1901, *Mon. Cul.*, 1: 268. *Pardomyia* Theobald, 1907, *Mon. Cul.*, 4: 280. *Ekrinomyia* Leicester, 1908, *Cul. of Malaya*, p. 71.

A rather small subgenus, species of which are distributed in the tropics of the Old World, from West Africa to the Solomon Islands, and in eastern Australia. The larvae of all species are predaceous on other mosquito larvae. One species is represented in Victoria.

Characters of the Subgenus

ADULT

Large species with conspicuous ornamentation on legs. Posterior pronotal bristles numerous. Lower mesepimeral bristles present. M_{3+4} arising proximally to r-m. Hind legs very long.

LARVA

Mouth parts modified for predacity. Ventral brush large. Lateral comb, a large patch of scales.

Aedes (Mucidus) alternans (Westwood)

Culex alternans Westwood, 1835, *Ann. Soc. ent. France*, 4: 681. *Culex commovens* Walker, 1856, *Ins. Saund. Dipt.*, 1: 432. *Culex hispidosus* Skuse, 1889, *Proc. Linn. Soc. N.S.W.*, 3: 1726. *Mucidus alternans* Edwards, 1923, *Bull. ent. Res.*, 14: 367. *Aedes alternans* Edwards 1932, *Gen. Insect.*, 194: 134.

This is a very large mosquito with a wing span of about 17 mm. It is readily recognized by the shaggy scaling of the legs and by the white bands on the femora, tibiae and tarsi.

ADULT FEMALE

Vertex with upright scales white medially, ochreous laterally. Eyes edged with white scales. Antennae ochreous, bases of segments dark. Proboscis mainly yellow scaled, becoming dark towards base and tip. Palps (fig. 27 d) half to two-thirds length of proboscis; clothed basally with ochreous and black scales with some white scales dorsally on first seg-

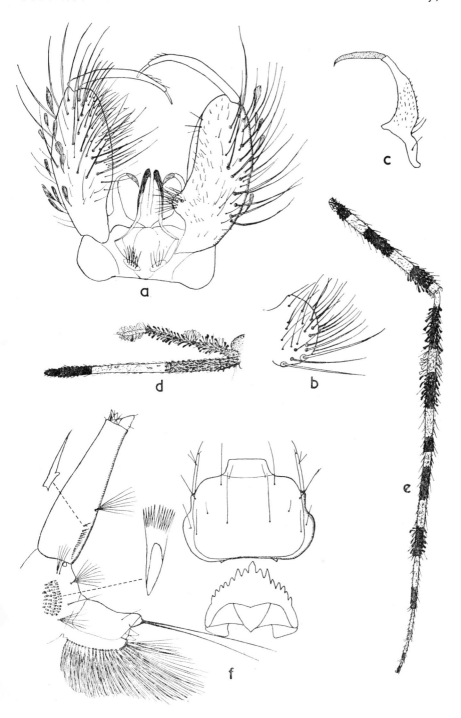

Fig. 27 *Aedes alternans* (Westwood). *a-e*, adult: *a*, male terminalia; *b*, basal lobe of coxite; *c*, harpago; *d*, female proboscis and palp; *e*, hind leg. *f*, larva: head, mentum and terminal segments.

ment, and at base of second; terminal segment mostly white scaled. Scutum clothed with brown and yellowish scales with some areas of white scales. Scutellum white scaled. Anterior pronotum, posterior pronotum and pleuron with white scales. Several lower mesepimeral bristles. Wing scales mainly dark with some ochreous and pale ones scattered or in patches. Legs (fig. 27 e) shaggily clad with dark and ochreous scales and with white rings on femora, tibiae and tarsi. Claws of all legs toothed. Tergites ochreous with median and lateral white patches. Sternites with dark and ochreous scales and patches of white scales.

ADULT MALE

Palps longer than proboscis for almost the length of terminal segment. *Terminalia* (fig. 27 a-c). Coxite about twice as long as broad; sternally with a number of long strong setae directed mesially. Basal lobe moderately large with numerous rather long setae. Style slightly shorter than coxite, appendage long. Harpago stout; appendage narrow, blade-like. Lobes of IXth tergite poorly defined, with 3-5 fine setae.

LARVA (fig. 27 f)

Larvae large with predacious mouth brushes. Antenna about half length of the head; seta 1 usually 2 branched. Head setae: 4-8 single; 9, single or 2 branched. Mentum with strong central tooth and 7-8 lateral teeth on each side. Prothoracic setae: 1, 2, 3, 5, 6 and 7, single; 4, 4-5 branched. *Abdomen.* VIIIth segment: lateral comb of about 50 fringed scales; seta 1, 13-17 branched; 2 and 4, single; 3, 5-9 branched; 5, 4-5 branched. Siphon with index 3·3-3·7; pecten on basal third of 17-23 spines; seta 1 about one-third length from base, 7-8 branched. Anal segment: saddle covering dorsal half of segment; setae 1, 2 and 3 single; 4 (ventral brush) of 27-30 tufts along whole length of segment. Anal papillae, equal, narrow, slightly less than length of saddle.

BIOLOGY

Ae. alternans breeds in brackish and fresh-water pools and swamps; from second stage the larvae are predacious. This is a day-biting mosquito which attacks man, domestic animals and birds.

DISTRIBUTION

In Victoria it is largely restricted to the northern part, particularly the Murray Valley. It is also recorded from New Guinea, New Caledonia, Northern and Eastern Australia.

Subgenus OCHLEROTATUS lynch Arribalzaga

Ochlerotatus Lynch Arribalzaga, 1891, *Rev. Mus. La Plata,* 1: 374; 2: 143.

For synonymy see Stone, Knight and Starcke, 1959.

Edwards (1924) divided the Australian species of the subgenus *Ochlerotatus* into two groups. In the first group he placed the species which have at least one lower mesepimeral bristle; in the second group he placed species without lower mesepimeral bristles.

Mackerras (1927) recognized Edwards' groups but he pointed out that the presence or absence of lower mesepimeral bristles is not constant in some species, e.g. *Aedes flavifrons* and *Aedes procax,* and indicated that the male terminalia present valuable characters for grouping the members of this subgenus.

Subsequently Edwards (1932) divided the *Ochlerotatus* species of the world into eight groups, of which three were represented in the Australian Region. The Australian species have since been divided by Marks (1957) into nine sections according to the structure of the male terminalia, and in some cases on larval characters also. Dobrotworsky (1960) pointed out that, on this system, two closely related species, *nivalis* Edw. and *silvestris* Dobrot., would fall into different sections. He considered that the creation of a Camptorhynchus Section was not justified and suggested that it should be merged into the Perkinsi Section, thus reducing the total of nine sections to eight. This modification has been accepted by Marks.

Five Sections, Burpengaryensis, Flavifrons, Perkinsi, Cunabulanus, and Stricklandi, house species with Holarctic affinities and are well represented in Victoria.

Characters of the Subgenus

Vertex and scutellum with narrow scales. Proboscis slender, longer than front femora. Male palpi as long as or longer than proboscis; last two segments hairy, projecting straight forwards or turned slightly downwards. Antennae of male with plume-hairs directed mainly dorsally and ventrally. Lower mesepimeral bristles present or absent. Fore claws always toothed in both sexes. VIIIth segment of female abdomen small and completely retractile; cerci long and narrow. *Male terminalia.* Coxite long, with distinct basal lobe and with more or less strongly developed apical lobe. Harpago well developed with flattened or setiform appendage. Style long, slender, with long terminal spine. Paraproct usually with single tooth. Phallosome simple, smooth, not divided into lateral lobes.

KEY TO VICTORIAN SPECIES OF THE SUBGENUS OCHLEROTATUS

Adult Females

1	Scutum creamy or white, with 4 longitudinal black stripes; tarsi banded, first segment with extensive creamy scaling	*vittiger*
	Scutum otherwise	2
2(1)	Tarsi with broad white bands	3
	Tarsi unbanded or with inconspicuous bands	7
3(2)	Wing entirely dark scaled	4
	Wing with some white scales	5
4(3)	Hind tarsal segment 5 with basal band; femora and tibiae mottled; proboscis always with some pale scales	*camptorhynchus*
	Hind tarsal segment 5 all white; femora and tibiae unmottled; proboscis entirely black	*calcariae*
5(3)	Lower mesepimeral bristles present; wing membrane with dark blotch; creamy scales on wings only slightly broader than dark ones; all hind tarsal segments banded	*flavifrons*

	Lower mesepimeral bristles absent 6
6(5)	Wings with numerous broad white scales; hind tarsal segment 5 entirely black; tergites with narrow or reduced basal bands .... *theobaldi*
	Wings with a few white scales the same size as dark ones and mainly along Sc and R; all hind tarsal segments with basal band; tergites with broad, straight basal bands .... *vigilax*
7(2)	Wing extensively mottled with pale scales 8
	Pale scales on wings if present usually only on C, Sc and R .. 9
8(7)	Tergites dark scaled; white scales on pleura contrasting with dark scutum .... *stricklandi*
	Tergites mottled .... *spilotus*
9(7)	Scutum with broad margin of pale scales contrasting with darker scales medially 10
	Scutal scaling otherwise 11
10(9)	Patch of broad flat scales in front of wing root; abdomen unbanded .... *andersoni*
	No patch of broad flat scales in front of wing roots; abdomen banded .... *nigrithorax*
11(9)	Fore and mid femora mottled 12
	All femora unmottled 18
12(11)	Scutal integument dark brown or almost black; tergites with basal white bands .... *sagax*
	Scutal integument reddish 13
13(12)	Wings with pale scales on distal part of C and R; tergites unbanded, IV-VII tergites mottled .... *luteifemur*
	Wing black scaled 14
14(13)	Hind femora pale on basal half or two-thirds 15
	Hind femora mottled 16
15(14)	Scutal scales small, dark bronze; tergites with white basal bands convex posteriorly .... *silvestris*
	Scutal scales rather large, golden with some admixture of bronze scales medially; tergal bands II-VI straight .... *nivalis*
16(14)	Scutum with some dark-bronze scales medially; proboscis black, tergites black .... (Grampians form) *andersoni*
	Scutum uniformly clothed with golden scales 17
17(16)	Proboscis black scaled .... *cunabulanus*
	Proboscis mottled or at least with some white scales on basal half .... *continentalis*
18(11)	Some broad scales on posterior pronotum; postspiracular area with broad scales; hind femur dark on apical quarter anteriorly and posteriorly; all or part of venter purplish .. *purpuriventris*
	Posterior pronotum with narrow scales only 19
19(18)	Scutal integument reddish, scales golden with admixture of black scales .... *clelandi*
	Scutal integument black or brownish-black; scutal scaling golden with two black stripes along dorsocentral bristles and two shorter stripes above wing roots .... *imperfectus*

VIGILAX SECTION

Male terminalia. Apical lobe of coxite undeveloped or weakly developed with one or two strong setae or fine hairs only; basal lobe rounded, hairy, sometimes with a few broadened setae apically, or produced into a curved arm with expanded knob bearing stout setae; no rows of long mesially directed setae sternally. Harpago slender, without thumb; appendage bristle-like, straight or slightly curved at tip.

Aedes (Ochlerotatus) vigilax (Skuse)

Culex vigilax Skuse, 1889, *Proc. Linn. Soc. N.S.W.*, (2), 3: 1731. *Culex albirostris* Macquart, 1850, *Dipt. Exot.*, Suppl. 4: 10. *Aedes vigilax* Edwards, 1922, *Bull. ent. Res.*, 13: 99.
For other synonyms consult Klein and Marks (1960).

A black mosquito with banded tarsi and abdomen. The proboscis is black except for a sharply-defined pale middle section of the underside. The wings are mottled with small white scales.

The type specimens on which the species *Culex albirostris* Macquart was described was re-examined recently (Klein and Marks, 1960). This examination disclosed that the type specimen is in fact an *Aedes* and that *Aedes vigilax* (Skuse) is synonymous with it. Application is being made to the International Commission of Zoological Nomenclature to validate the specific name *vigilax* Skuse, 1889, and reject the specific name *albirostris* Macquart, 1850. A change in the name of this common species would result in much confusion among health workers.

ADULT FEMALE

Vertex with narrow pale scales; upright forked scales black. Proboscis (fig. 28 c) black scaled with a sharply defined pale area in the middle on the underside. Palps dark with white patch at tip. Scutal integument almost black; scutal scales mainly dark bronze, with admixture of some golden scales. Posterior pronotum clothed with elongate dark scales and with patch of white scales below. Postspiracular area with patch of white scales. No lower mesepimeral bristles. Wings with scattered white scales of normal size mainly along anterior border. Legs black, mottled (fig. 28 d); tarsi banded; fifth hind tarsal segment with white band. Tergites black with broad white basal bands with straight posterior edge. Sternites white scaled with black apical bands or large lateral apical patches of black scales.

ADULT MALE

Palps black, slightly shorter than proboscis; all segments with patches of white scales at base. *Terminalia* (fig. 28 a, b). Coxite short, about 2½ times as long as broad. Apical lobe very small with two strong moderately long setae. Basal lobe prominent with numerous fine, short setae and some rather strong ones with curved tips. Harpago short, appendage seta-like, almost as long as harpago. Paraproct with single tooth. Phallosome simple. Lobes of IXth tergite with 4-6 strong setae.

LARVA (fig. 28 f)

Head, siphon and saddle light brown. Head about two-thirds as long as broad. Antenna about half length of head; seta 1, 3-4 branched. Head setae: 4, tiny, 2-3 branched; 5 and 6, single; 7, 5-7 branched; 8 and 9, 2-3 branched. Prothoracic setae: 1 and 2, single; 3, 2-3 branched; 4, 3 branched; 5 and 7, 2 branched; 6, single. *Abdomen.* VIIIth segment: lateral comb 19-33 fringed scales; seta 1, small, 2-4 branched; 2 and 4,

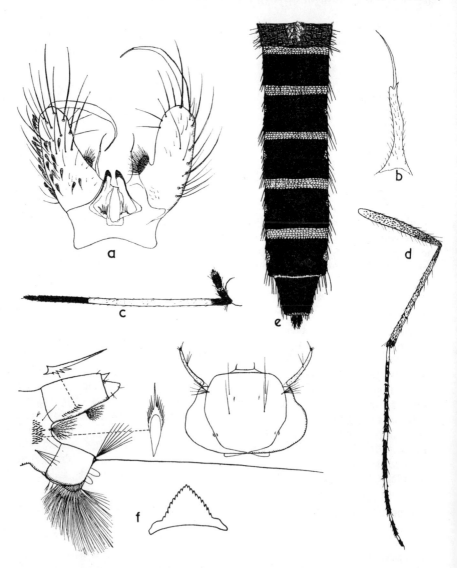

FIG. 28 *Aedes vigilax* (Skuse). *a-e*, adult: *a*, male terminalia; *b*, harpago; *c*, proboscis and palp of female; *d*, hind leg; *e*, female abdomen, dorsal view. *f*, larva: head, mentum and terminal segments.

single; 3, 6-12 branched; 5, 3-4 branched. Siphon short, slightly tapering, index 1·7-2·3; pecten of 6-11 spines with three or four small denticles at base; seta 1, 8-11 branched, arising about half way along siphon. Anal segment: saddle covering about seven-eighths of dorsal part of segment; setae 1 and 3, single; 2, 6-9 branched; 4 (ventral brush) of 14-16 tufts, two usually precratal. Anal papillae almost equal, slightly less than length of saddle.

BIOLOGY

Ae. vigilax breeds in brackish water, salt marshes and also in fresh-water ground pools.

DISTRIBUTION

Ae. vigilax is typically a coastal species, but in Victoria is restricted to the Mildura area; it has been recorded from South Australia and Western Australia (Perth) and is widely distributed along the coastal areas of northern Australia, New Guinea and the Oriental Region.

Aedes (Ochlerotatus) procax Skuse

Culex procax Skuse, 1889, *Proc. Linn. Soc. N.S.W.*, 3: 1742. *Aedes (Ochlerotatus) procax*, Klein and Marks, 1960, *Proc. Linn. Soc. N.S.W.*, 85: 112.

In its general appearance this species resembles the common *Ae. rubrithorax* but is distinguished from it by the lack of mottling on the legs, by its pointed abdomen and by the black apical bands on the sternites.

ADULT FEMALE

Vertex with narrow curved light-golden scales; upright forked scales dark. Proboscis dark scaled, pale in middle on underside. Palps dark scaled. Scutal integument brown. Scutal scales golden. Posterior pronotum clothed with elongate dark scales. Postspiracular area without scales. Lower mesepimeral bristles absent. Wing dark scaled. Legs (fig. 29 e) dark scaled, without mottling. Hind femora pale on basal three-quarters. Tarsi with white basal bands; fifth hind tarsal segment with white band. Tergites black scaled with basal white bands on segments 2-7, bands on segments 2-4 usually constricted laterally. Large white lateral patches on all segments. Sternites white scaled with black apical bands sometimes broken in middle.

ADULT MALE

Proboscis dark scaled. Palps longer than proboscis with labella, dark scaled. *Terminalia* (fig. 29 a-c). Coxite short, with scales and long setae laterally; tergally with a few small setae. Apical lobe absent; basal lobe prominent with numerous moderately long setae. Harpago slender with straight bristle-like appendage. Phallosome simple, tubular. Lobes of IXth tergite with 5-7 strong setae.

LARVA (fig. 29 f)

Head, siphon, and saddle light brown. Head about two-thirds as long as broad. Antenna about half length of head; spiculated; setae 1, single or 2 branched. Head setae: 4, single; 5, 2-3 branched; 6, single; 7, 5-6 branched; 8 and 9, 3-4 branched. Prothoracic setae: 1, 5 and 6, single; 2, 2-3 branched; 3 and 7, 2 branched; 4, 2-4 branched. *Abdomen.* VIIIth segment: lateral comb of about 25 scales; seta 1 and 2, single or 2 branched; 3, 9-10 branched; 4, single; 5, 5-7 branched. Siphon short, index about 2·7; pecten of 11-13 unidentate spines; seta 1, 7-10 branched. Anal segment: seta 1 and 3 single; 2, 12-13 branched, 4 (ventral brush)

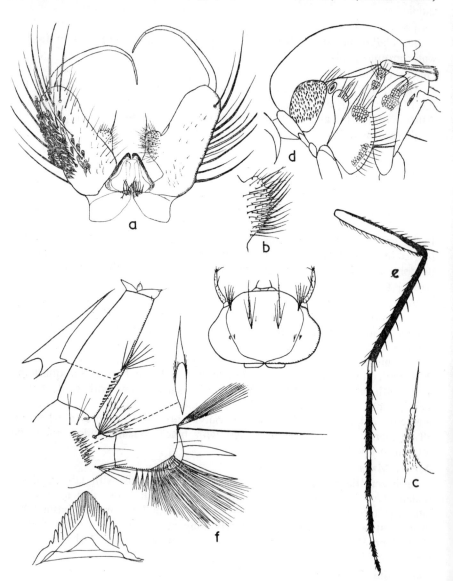

Fig. 29 *Aedes procax* Skuse. *a-e*, adult: *a*, male terminalia; *b*, basal lobe; *c*, harpago; *d*, side view of thorax; *e*, hind leg. *f*, larva: head, mentum and terminal segments.

of 16-17 tufts, usually 3 precratal. Saddle almost complete ring. Anal papillae almost equal, slightly less than length of saddle, narrow, pointed.

BIOLOGY AND DISTRIBUTION

A rare species in Victoria. Only one specimen, a female, has been collected near Genoa (East Gippsland). It also occurs in Queensland and New South Wales.

THEOBALDI SECTION

Male terminalia. Apical lobe of coxite elongate, fairly well developed, hairy; basal lobe prominent, densely hairy often with some modified setae; no rows of long mesially directed setae sternally. Harpago slender without thumb; appendage bristle-like, angled near tip, and may have retrorse projection.

LARVA

Lateral comb of 12-25 spines with denticles or fringe at base. Siphon not sclerotized to base dorsally; pecten without detached teeth; seta 1 long, arising beyond pecten; seta 9 a stout hook.

Aedes (Ochlerotatus) theobaldi (Taylor)

Grabhamia flavifrons Theobald, 1907, (*non* Skuse), *Mon. Cul.*, 4: 304. *Grabhamia theobaldi* Taylor, 1913, *Proc. Linn. Soc. N.S.W.*, 38: 751, ibid 1919, 43: 832. *Aedes theobaldi* Edwards, 1924, *Bull. ent. Res.*, 14: 375. Marks, 1949, 2: 11.

This species is similar to *Ae. vigilax* but the white scales on the wings are far more numerous and are broad, not narrow. The pale area on the underside of the proboscis has ill-defined borders.

ADULT FEMALE

Vertex with pale curved and upright forked scales, latter becoming dark laterally and towards neck. Proboscis black scaled; pale area in middle on underside with ill-defined borders (fig. 30 c). Palps dark scaled with some admixture of white scales. Posterior pronotum with elongate white scales below; above with elongate and narrow dark scales, with or without some white ones. Scutal integument dark brown. Scutum clothed with narrow, golden and creamy scales with some admixture of dark-bronze ones. Wings (fig. 30 d) with extensive mottling of large white scales. Femora, tibiae and first tarsal segment of hind legs mottled; tarsi banded. In Victorian specimens tarsal segment 5 always without a basal white band (fig. 30 e). Tergites (fig. 30 f) black scaled, sometimes mottled and with straight or constricted basal bands, which may be reduced, and lateral patches. Sternites white scaled with apical black bands or lateral patches of black scales or with mottling of black scales.

ADULT MALE

Palps longer than proboscis, dark scaled, with patch of white scales at base of all segments and some scattered white scales along them. *Terminalia* (fig. 30 a, b). Coxite about four times as long as broad. Apical lobe

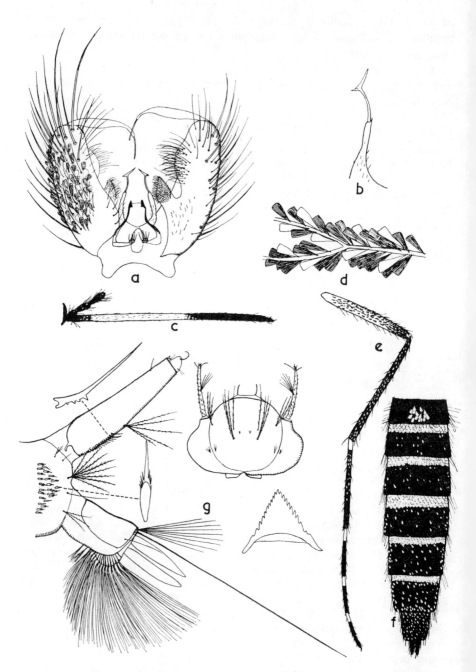

Fig. 30 *Aedes theobaldi* (Taylor). *a-f*, adult: *a*, male terminalia; *b*, harpago; *c*, proboscis and palp of female; *d*, portion of wing showing scales; *e*, hind leg; *f*, female abdomen, dorsal view. *g*, larva: head, mentum and terminal segments.

with numerous straight mesially directed setae; basal lobe prominent, broadening apically and clothed all over with fine short setae; setae along apical margin rather stout but showing no special modification. Style narrow at base and widening towards middle with apical quarter curved. Harpago slender, appendage long, slender, angled near tip; angle produced in retrorse projection. Paraproct with single tooth. Phallosome simple, oval. Lobes of IXth tergite with 4-5 setae.

LARVA (fig. 30 g)

Pale with light brown head, siphon and saddle. Head two-thirds as long as broad. Antenna about two-thirds length of head, curved, spiculate; seta 1, 4-7 branched. Head setae: 4, minute tuft; 5, 2-3 branched; 6, single; 7, 2-3 branched; 8, 2-3 branched; 9, small tuft. Prothoracic setae: 1 and 2, single; 3, 1-3 branched; 4 and 6, single; 5, 2 branched; 7, 3 branched. *Abdomen.* VIIIth segment: lateral comb of 17-24 pointed spines arranged in 2-3 rows, spines coarsely fringed basally; seta 1, 2-6 branched; 2 and 4, single; 3, 6-11 branched, plumose; 5, 2-5 branched. Siphon slightly tapering beyond tuft; sclerotization not extending to base dorsally; tracheae very narrow; index 3·3-3·9; pecten of 9-15 spines, each with 2-5 denticles at base; seta 1, long, plumose, 4-9 branched, arising a third to two-thirds along length of siphon. Anal segment: saddle covering about two-thirds dorsal part of segment; setae 1 and 3 single; 2, 7-13 branched; 4 (ventral brush) of 15-17 tufts, 1-2 precratal. Anal papillae equal, pointed, twice as long as saddle.

BIOLOGY

Ae. theobaldi breeds in clean or cloudy pools and ditches with grassy edges. The larvae are mostly found lying on their backs on the bottom. *Ae. theobaldi* may have several generations a year; the first spring generation usually develops from eggs which lie dormant during the summer. In irrigated areas such as Mildura the periodic flooding of breeding places permits a number of summer generations.

It is a day-biting species, which attacks man and domestic animals.

DISTRIBUTION

In Victoria it is widely distributed north of the Central Highlands, but odd specimens have been collected in the south. This species has also been recorded from Queensland, New South Wales and South Australia where again it is more common in inland areas.

Aedes (Ochlerotatus) theobaldi eidsvoldensis Mackerras

Aedes theobaldi form *eidsvoldensis* Mackerras, 1927, *Proc. Linn. Soc. N.S.W.*, 52: 295. *Aedes theobaldi eidsvoldensis* Marks, 1949, *Pap. Dep. Ent. Univ. Qd,* 2 (11): 17.

The status of this subspecies is uncertain. It is known only from the female, which can be distinguished from the typical form by the narrower pale scales of the wings, by the dull creamy golden scaling of the scutum, and by the fact that the pale bands of the tergites are produced posteriorly into a prominent triangle.

DISTRIBUTION

Only a single specimen has been collected in Victoria (Mildura area) but it is known from several localities in Queensland.

BURPENGARYENSIS SECTION

Male terminalia. Apical lobe of coxite slightly developed, with 1-2 strong setae; basal lobe moderately developed with numerous setae on inner fold and one long seta at apex; no rows of long mesially directed setae sternally. Harpago with prominent thumb with apical bristle; appendage broad, fimbriated.

LARVA

Lateral comb 6-10 spines. Pecten with detached teeth beyond siphonal seta 1.

Aedes (Ochlerotatus) nigrithorax (Macquart)

Culex nigrithorax Macquart, 1847, *Dipt. Exot.* Suppl. 2: 9.
Aedes nigrithorax Edwards, 1924, *Bull. ent. Res.*, 14: 377.
Aedes macleayanus Mackerras, 1927, *Proc. Linn. Soc. N.S.W.*, 52: 291. *Aede nigrithorax* Marks, 1960, *Proc. Linn. Soc. N.S.W.*, 85: 117-20.

The scutum dark bronze mesially and creamy, or white, laterally. The fore and mid femora are mottled but the hind are pale almost to the apex; the tarsi are unbanded.

ADULT FEMALE

Vertex clothed with narrow curved whitish scales; upright forked scales creamy. Proboscis and palps dark scaled. Scutal integument black. Scutum (fig. 31 c) clothed with narrow dark bronze scales between dorsocentral bristles; this bronze area widens behind middle of thorax; scutum clothed laterally with larger curved creamy or white scales. Posterior pronotum clothed with narrow bronze scales above and broader elongate scales below. Usually four lower mesepimeral bristles. Wing dark scaled; some white scales on base at costa. Fore and mid femora mottled; hind femora usually pale almost to apex, anteriorly, with apicodorsal dark scaling (fig. 31 d). All tibiae and basal half of first tarsal segment extensively mottled. Tarsi unbanded. Claws equal with a strong tooth; or one or both hind claws simple. Tergites (fig. 31 e) black scaled, with creamy basal bands joining lateral patches of white scales; these bands may be narrow or broad, almost parallel or widened medially; tergite VII with basal band and some mottling medially or all creamy scaled. Sternites creamy scaled with small apical lateral patches of black scales and sometimes a few scattered black scales.

ADULT MALE

Palps as long as proboscis. *Terminalia* (fig. 31 a, b). Coxite about 3 times as long as broad with fairly well developed apical and basal lobes. Apical lobe pointed with 2-3 long flattened and several small setae. Basal lobe with large patch of fairly short fine setae and a single long seta on lower tergal edge of lobe. Style slender, curved on apical third. Harpago

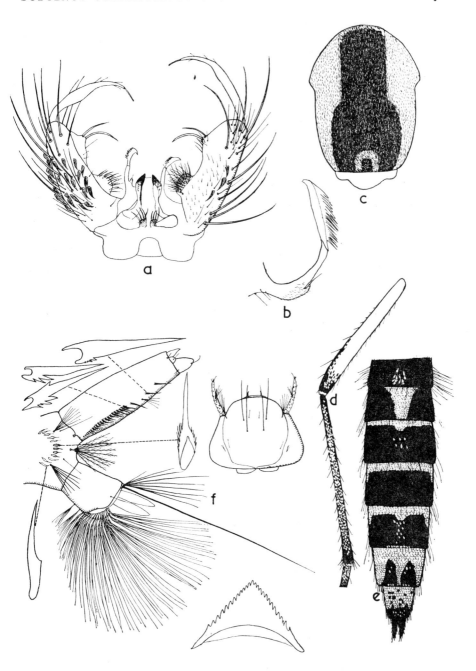

FIG. 31 *Aedes nigrithorax* (Macquart). *a-e*, adult: *a*, male terminalia; *b*, harpago; *c*, thorax; *d*, hind leg; *e*, female abdomen, dorsal view. *f*, larva: head, mentum and terminal segments.

stout, long, curved, with sub-basal thumb bearing a moderately long seta at tip; appendage long, fimbriated. Paraproct with single tooth. Lobes of IXth tergite with 5-8 stout setae.

LARVA (fig. 31 f)

Head, siphon and saddle light brown. Head about two-thirds as long as broad. Antenna about half length of head, spiculate; seta 1 arising about half length of antenna, 5-6 branched. Head setae: 4, 2-3 branched; 5, single or 2 branched; 6, single; 7, 6-8 branched; 8, single; 9, 2 branched. Prothoracic setae: 1, 2, 4 and 6, single; 3, single or 2 branched; 5, usually 2 branched, may be single; 7, 3-4 branched. *Abdomen.* VIIIth segment: lateral comb row of 8-14 spines coarsely fringed basally; seta 1, 3-6 branched; 2 and 4, single; 3, 6-11 branched; 5, 5-9 branched. Siphon slightly tapering towards apex; index 2·8-3·6, mean 3·0; pecten of 19-28 spines, 1 or 2 detached spines beyond seta 1; seta 1, 3-4 branched arising above mid length of siphon. Anal segment: saddle covering dorsal seven-eighths of segment; seta 1 and 3, single; 2, 7-9 branched; 4 (ventral tuft) of 16-17 tufts, 1-2 precratal. Anal papillae lanceolate, about as long as or longer than the saddle.

BIOLOGY

In Victoria *Ae. nigrithorax* is largely confined to wooded plains or shallow valleys in slightly elevated country; it is most abundant in sparse bushland and usually does not penetrate deeply into dense forests or mountains. It breeds in more or less shallow pools that have decayed leaves on the bottom and are exposed to the sun for the greater part of the day, or lit by dispersed light. Sometimes larvae are found in deeper pools, small dams or swamps.

The seasonal appearance of adults depends on the filling of breeding places by the autumn-winter rains. Adults have been collected in early September, but a dry autumn and early winter may delay hatching and growth of larvae, so that the adults appear a month or so late. A further generation is possible when heavy summer rains fill breeding places.

Ae. nigrithorax is a day-biting species which is not usually attracted to man in large numbers, but in certain places near its breeding sites, and where it is numerous, it attacks very persistently.

DISTRIBUTION

In Victoria *Ae. nigrithorax* is distributed on and south of the Central Highlands. It occurs also in New South Wales, South Australia and Tasmania.

Aedes (Ochlerotatus) sagax (Skuse)

Culex sagax Skuse, 1889, *Proc. Linn. Soc. N.S.W.*, 3: 1744.
Culicada wilsoni Taylor, 1919, *Proc. Linn. Soc. N.S.W.*, 43: 833. *Aedes sagax* Edwards, 1924, *Bull. ent. Res.*, 14: 378.

The main distinguishing feature from *Ae. nigrithorax* is that the scutum is more or less uniformly clothed with bronze scales and has no sharply defined pale margins. The basal portion of the hind femora is mottled.

Adult Female

Ae. sagax is similar to *Ae. nigrithorax*. The main differences from *nigrithorax* are: Palps may be slightly mottled. Bronze scales on scutum not restricted to the area between dorsocentral bristles but spread to the fossa so that the margin of creamy-white scales becomes inconspicuous. 3-8 lower mesepimeral bristles. There are more white scales on the basal part of costa and such scales may be on Sc. and on base of R_1. Pale part of hind femora with variable number of dark scales (fig. 32 e). Claws equal; fore and mid claws usually with a strong tooth; hind claws with single tooth or simple. Tergites (fig. 32 d) sometimes have broader basal bands with apical border mesially broadened into an angle which may touch apical border of segment and make a longitudinal medial line on tergites.

Adult Male

Palps longer than proboscis, third segment with admixture of pale scales dorsally on basal half or two-thirds. Lower mesepimeral bristles absent. Terminalia (fig. 32 a-c) identical with those of *nigrithorax*.

Larva (fig. 32 f)

The larvae of *Ae. sagax* are almost identical with those of *Ae. nigrithorax* and the two species cannot be distinguished with certainty. The following morphological traits distinguish *Ae. sagax*: Seta 1 of VIIIth abdominal segment, 4-8 branched; 3, 7-14 branched. Siphon index 2·3-2·8, mean 2·6; pecten of 20-25 spines, mean 23; usually only one detached tooth above seta 1. Anal segment: seta 2, 9-12 branched.

Biology

Ae. sagax is a northern species and is confined to more or less flat country. It breeds in the open, but it is more common in and near woodland. The larvae have been collected in a variety of habitats; in more or less shallow pools exposed to the sun, and with a grassy bottom and clean water; in roadside ditches with grassy edges; in old irrigation channels overgrown with reeds; and in muddy water in shallow mining pits, with decayed leaves on bottom but no vegetation.

In non-irrigated areas *sagax* is a typical spring species and usually has only one generation a year, but if the autumn rains start early two winter-spring generations are possible. Adults of the first generation have been collected in mid-June. In irrigated areas it has several generations, the number of which depends on flooding and drying out of breeding sites. In the Mildura area adults have been collected from November to April inclusive.

Ae. sagax is a day-biting mosquito which has been recorded attacking man and domestic animals.

Distribution

Ae. sagax is distributed mainly north of the Central Highlands in Victoria but its distribution overlaps that of *nigrithorax* particularly in the Western District. It has not been collected in Gippsland. It has also been recorded in New South Wales, South Australia and Western Australia.

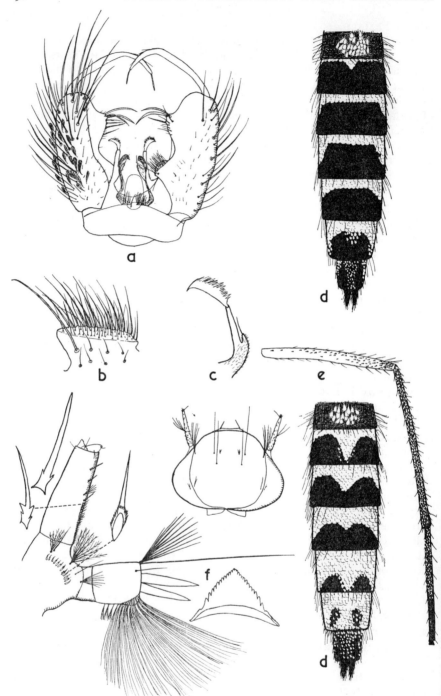

FIG. 32 *Aedes sagax* (Skuse). *a-e*, adult: *a*, male terminalia; *b*, basal lobe of coxite; *c*, harpago; *d*, variations in ornamentation of female abdomen, dorsal view; *e*, femur, tibia and first tarsal segment of hind leg. *f*, larva: head, mentum and terminal segments.

Aedes (Ochlerotatus) vittiger (Skuse)

Culex vittiger Skuse, 1889, *Proc. Linn. Soc. N.S.W.*, 3: 1728. *Aedes vittiger* Edwards, 1923, *Bull. ent. Res.*, 14: 372.

A large mosquito which is easily distinguished from all other species by the presence of four sharply defined lines of small black scales on the scutum, which is elsewhere clothed with ochreous-white scales.

ADULT FEMALE

Vertex clothed with narrow curved and upright forked creamy scales. The latter may become dark laterally. Proboscis black basally and apically, extensively mottled on middle three-quarters; alternatively it may be mottled at base and apically and creamy scaled between. Palps mottled on basal three-quarters, dark apically. Scutum (fig. 33 e) clothed with narrow ochreous scales medially, with white scales laterally and with four sharply defined lines of smaller black scales. Posterior pronotum clothed mostly with narrow pale scales; some broad and elongate scales below. Usually 3 lower mesepimeral bristles. Wing dark scaled with pale scales along C and Sc, and sometimes on base of R_1. Fore and mid femora and tibiae more or less mottled; hind femora pale-ochreous scaled on basal half or three-quarters; black or mottled dorsally on apical quarter; hind tibiae violet-black, mottled with creamy scales or creamy mottled with black scales. First tarsal segments pale scaled on basal half or three-quarters; hind tarsal segment 5 black. Tergites II-IV dark scaled with broad basal bands and sometimes mottled apically; V-VII ochreous-white scaled, sometimes with some dark scales on V-VI. Venter has more or fewer black scales along apical border of segments.

ADULT MALE

Palps longer than proboscis by about half length of fifth segment; violet-black with some pale scales on second segment. Legs violet-black with some mottling; tarsi may have only faint basal bands. *Terminalia.* Coxite 3-3½ times as long as broad. Apical lobe small, pointed, with one strong bristle and several small ones. Basal lobe moderate with numerous short fine setae and one long fine seta on lower tergal edge. Style curved, tapering toward apex. Harpago stout with subbasal thumb bearing long seta at tip; appendage fringed. Paraproct with single tooth. Lobes of IXth tergite with 6-7 short strong setae.

LARVA (fig. 33 f)

Head and siphon light brown. Antenna short; seta 1, 4-5 branched. Head setae: 4, 2 branched; 5, 6, and 8, single; 7, 5-6 branched; 9, single or 2 branched. Prothoracic setae: 1, 2, 3, 4 and 5, single; 6, single or 2 branched; 7, 3 branched. *Abdomen.* VIIIth segment: lateral comb single row of about 10 teeth; seta 1, 7-9 branched; 2 and 4, single, 3, 8-9 branched; 5, 3-5 branched. Siphon stout with index about 2·5; seta 1, single or 2-3 branched; pecten of 20-21 teeth, usually 2 or 3 teeth beyond siphonal seta 1. Anal segment: saddle almost complete ring; seta 1 and 3, single; 2, 6-7 branched; 4 of 19-20 tufts. Anal papillae about twice as long as saddle, narrow and pointed.

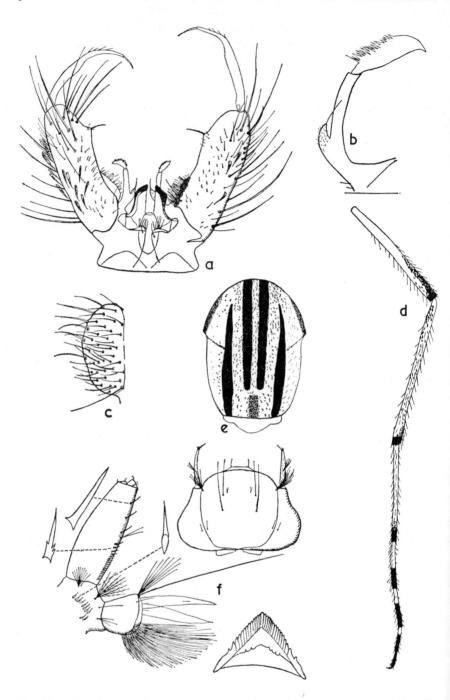

Fig. 33 *Aedes vittiger* (Skuse). *a-e*, adult: *a*, male terminalia; *b*, harpago; *c*, basal lobe of coxite; *d*, hind leg. *e*, thorax. *f*, larva: head, mentum and terminal segments.

Biology

Ae. vittiger breeds in clear or cloudy pools 2-4 feet deep and exposed to the sun; the bottom and edges are usually grassed. It has several generations a year; the first generation appears in September, but it is not a large one. The number of adults increases towards mid-summer and then declines by the end of autumn. *Ae. vittiger* is particularly numerous during mid-summer in the citrus orchards of the Mildura area where lucerne is grown. Apparently the dense and tall lucerne provides the most suitable resting places for the adults during the hot, dry days.

It is a very vicious day-biting mosquito which attacks man and domestic animals.

Distribution

Ae. vittiger occurs in the northern part of Victoria. It is particularly numerous in the Murray Valley, but it has also been collected well into the Central Highlands. Freshly emerged females, which had evidently bred locally, have been collected at Woods Point. It also occurs in Queensland, New South Wales and South Australia.

Aedes (Ochlerotatus) imperfectus Dobrotworsky

Aedes imperfectus Dobrotworsky, 1962, *Proc. Linn. Soc. N.S.W.*, 87: 296.

The scutum is clothed with golden, or brownish, scales with two narrow stripes of fine black scales. The legs are unmottled; the tarsi are unbanded. The tergites are unbanded with large white lateral patches.

Adult Female

Vertex clothed with narrow, and upright forked, creamy scales. Proboscis and palps black. Scutal integument black. Scutum clothed with golden or brownish scales with two stripes of fine black scales between acrostichal and dorsocentral bristles; some larger, curved pale scales in front of wing root. Posterior pronotum clothed with narrow golden scales above and elongate whitish scales below. Wing dark scaled. Fore and mid femora black anteriorly, pale posteriorly, unmottled. Hind femora (fig. 34 c) white on basal three-quarters, black scales at tip making a ring, broken below. Tibiae and tarsi violet-black. Tergites (fig. 34 d) violet-black scaled with large lateral basal patches; narrow basal bands may be present. Venter white scaled with median apical patches of black or ochreous scales.

Adult Male

Palps longer than proboscis, black scaled with a few pale scales dorsally on segments 1 and 2, usually 1-2 lower mesepimeral bristles, sometimes none. *Terminalia* (fig. 34 a, b). Coxite about 2½ times as long as broad. Apical lobe small, triangular, usually with one long curved flattened seta and several small ones. Basal lobe prominent, with patch of fairly short setae. Harpago stout, long, curved, with sub-basal thumb bearing a moderately long seta at tip; appendage long, fimbriated. Paraproct with single tooth. Lobes of IXth tergite with 3-6 short strong setae.

LARVA (fig. 34 f)

Head about five-sevenths as long as broad. Antenna about half length of head: seta 1, 5-7 branched, arising slightly below mid length of antenna. Head setae: 4, 2-3 branched; 5 and 6, single; 7, 4-7 branched; 8, single; 9, 2-3 branched. Prothoracic setae: 1, 2, 4, 5 and 6, single; 3, single or 2 branched; 7, 3-4 branched. *Abdomen.* VIIIth segment: lateral

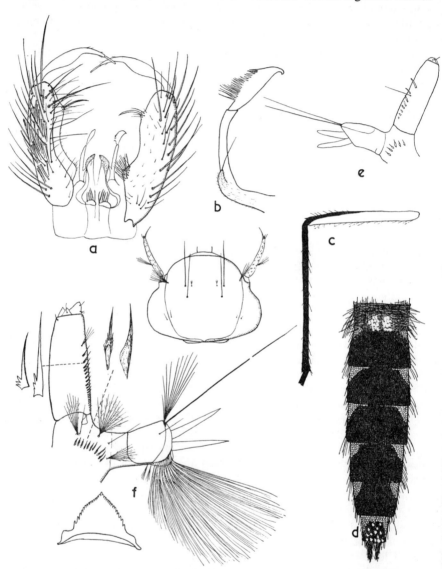

FIG. 34 *Aedes imperfectus* Dobrotworsky. *a-d*, adult: *a*, male terminalia; *b*, harpago; *c*, femur and tibia of hind leg; *d*, female abdomen, dorsal view. *e-f*, larva: *e*, terminal segments of first stage; *f*, head, mentum and terminal segments. (*a, b, e, f*, after Dobrotworsky.)

comb of 10-13 spines; seta 1, 4-7 branched; 2 and 4, single; 3, 6-10 branched, plumose; 5, 6-7 branched. Siphon moderately long, almost cylindrical, index 2·7-3·1; pecten of 23-25 spines; seta 1, 3-6 branched, arising two-thirds along siphon. Anal segment: saddle forming almost complete ring, a small separate sclerotized plate attached near proximal angle of saddle; setae 1 and 3, single; 2, 7-8 branched; 4 (ventral brush) of 14-16 tufts, 2-4 precratal. Anal papillae, pointed, about as long as saddle. The first-stage larva (fig. 34 e) has detached pecten teeth beyond seta 1.

BIOLOGY

Ae. imperfectus breeds in grassy pools which are usually filled by flood waters and shaded by trees for part of the day. It is a day-biting species which near breeding sites attacks man.

DISTRIBUTION

In Victoria it occurs throughout Gippsland; only single specimens have been collected west of it (Lyonville and Beaufort). It has also been recorded from Queensland and New South Wales.

FLAVIFRONS SECTION

Male terminalia. Apical lobe of coxite a moderate swelling, bearing short setae; basal lobe slightly developed with a patch of fine setae on inner fold; several rows of long mesially directed setae on apical half sternally. Harpago may have distinct thumb; appendage broad, fimbriated.

Aedes (Ochlerotatus) flavifrons (Skuse)

Culex flavifrons Skuse, 1889, *Proc. Linn. Soc. N.S.W.,* (2) 3: 1735. *Culicada vandema* Strickland, 1911, *Entom.,* 44: 202. *Culicada flavifrons* Taylor, 1914, ibid, 38: 751. *Aedes flavifrons* Edwards, 1924, *Bull. ent. Res.,* 14: 374. Dobrotworsky, 1960b, 85: 180.

Easily distinguished by the presence of a dark blotch on the wing membrane, the mottled wings and legs, and banded tarsi.

ADULT FEMALE

Vertex clothed with narrow, pale, and upright forked, dark or bronzed, scales. Proboscis and palps mottled. Scutal integument from light brown to dark brown. Scutum clothed with narrow pale-golden scales with some admixture of black scales; in dark specimens black scutal scales predominate over pale golden ones; black scales may form large patch in fossa. Mesepimeron with 2-5 lower bristles. Wing membrane (fig. 35 d) with dark blotch in middle. Wings mottled with white scales of the same size as the dark ones or only slightly larger. Femora and tibiae (fig. 35 e) mottled; tarsi banded; tarsal segment 5 of hind legs with basal white band. Tergites black scaled, unbanded; last segments mottled with ochreous scales. Venter either ochreous, mottled with black scales, or basal part black scaled with some ochreous mottling and apical segments ochreous.

ADULT MALE

Palps longer than proboscis, dark scaled with patches of pale scales at

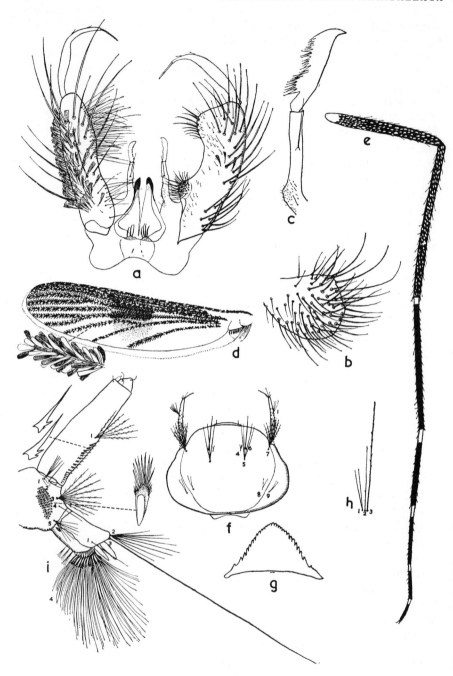

FIG. 35 *Aedes flavifrons* (Skuse). *a-e*, adult: *a*, male terminalia; *b*, basal lobe of coxite; *c*, harpago; *d*, wing; *e*, hind leg; *f-i*, larva: *f*, head; *g*, mentum; *h*, prothoracic setae 1, 2 and 3; *i*, terminal segments. (*a-c, i-h*, after Dobrotworsky.)

base of segments 4 and 5. Lower mesepimeral bristles absent. *Terminalia* (fig. 35 a-c). Coxite 4 times as long as broad with dense, long, mesially directed, yellowish setae arising sternally on distal half. Apical lobe moderate, with several fine short setae and some stronger, longer ones. Basal lobe a rounded plate with numerous fine setae, and 2-3 longer ones arising on tergal side. Style curved, swollen at about mid length, apical third slender. Harpago stout, curved with one strong, long preapical seta; appendage almost as long as harpago, fimbriated and with curved tip. Paraproct with one tooth. Lobes of IXth tergite with 3-6 strong setae.

LARVA (fig. 35 f-i)

Head, siphon and saddle light brown. Head about two-thirds long as broad. Antenna about half length of head, spiculate; seta 1 arising at about two-thirds length from base, 3-5 branched. Head setae: 4, tiny, 2-4 branched; 5, 3-4 branched; 6, single; 7, 8-10 branched; 8 and 9, 2 branched. Mentum with 13-14 lateral teeth on each side. Prothoracic setae: 3, 2 branched or single; 7, 2-3 branched; all others single. *Abdomen*. VIIIth segment: lateral comb patch of 36-53 coarsely fringed scales; seta 1, 4-5 branched; 2 and 4, single; 3, 8-10 branched; 5, 6-7 branched. Siphon slightly tapering towards apex, index 3·0-3·6; pecten of 15-18 slender spines; seta 1, 5-6 branched, arising half-way along siphon. Anal segment: saddle covering dorsal two-thirds of segment; a small separate sclerotized plate lies near proximal angle of saddle; setae 1 and 3, single; 2, 7-8 branched; 4 (ventral brush) of 16 tufts, 3 usually precratal. Anal papillae pointed, about one-third length of saddle.

BIOLOGY

Ae. flavifrons is one of the most common spring mosquitoes in southern Victoria. It breeds in ground pools, up to two feet in depth, containing leaves and twigs, and usually without vegetation. The pools may be exposed to the sun for a part of the day or completely shaded; pits, sometimes 2-3 feet deep, at the base of uprooted trees are favoured breeding sites.

In Victoria *flavifrons* is confined to woodland at low altitudes; it has not been collected above 2000-3000 feet. It is a day-biting mosquito, and attacks man; it remains active in the shade during the day even in the summer.

DISTRIBUTION

Ae. flavifrons is very common on and south of the Central Highlands in Victoria. It is also recorded from New South Wales, South Australia and Tasmania.

Aedes (Ochlerotatus) calcariae Marks

Aedes calcariae Marks, 1957, *Pap. Dep. Ent. Univ. Qd*, 1 (5): 74.

Easily distinguished from all other species by the unmottled legs and wings, by the banded tarsi and by the fifth tarsal segment of the hind leg which is entirely white.

Adult Female

Vertex clothed with narrow pale and upright forked golden scales. Proboscis and palps black scaled; last segment of palps with patch of white scales on tip. Scutal integument brown. Scutum with narrow bronze-brown scales; indistinct pattern of creamy scales predominantly along border, along dorsocentral and acrostichal bristles; creamy scales dominant in fossa. Posterior pronotum clothed with narrow curved, and some elongate, dark and pale scales. Wings black scaled. Legs (fig. 36 d) black scaled, unmottled; tarsi with white basal bands, basal bands on hind tarsi extending onto apices of preceding segments; segment 5 all white. Tergites purplish-black, unbanded or with narrow basal bands, and large basal lateral white patches. Sternites pale scaled with basal median, and apical lateral, patches of black scales.

Adult Male

Palps about as long as proboscis, purplish-black scaled with dorsal white patches at base of segments 4 and 5. Tergites with broad basal white bands. *Terminalia* (fig. 36 a-c). Coxite about 4 times as long as broad. Sternally with numerous long golden, mesially directed setae. Apical lobe moderate with 10-16 short setae. Basal lobe extended as round plate with numerous fine setae and 2-3 longer ones. Style about half length of coxite, curved; apical third tapering. Harpago curved, long, stout, with one strong preapical seta; appendage almost as long as harpago, blade-like, fimbriated, with curved, pointed tip. Paraproct with 1 strong and 2 inconspicuous teeth. Phallosome simple, oval. Lobes of IXth tergite with 2-5 strong short setae.

Larva (fig. 36 e)

Head, siphon and saddle brown. Head four-fifths as long as broad. Antenna almost half length of head; seta 1 arising about mid length, 3-6 branched. Head setae: 4, 4-7 branched; 5, single or 2-5 branched; 6, single or rarely 2-3 branched; 7, 6-8 branched; 8, single or rarely 2-3 branched; 9, 2-4 branched. Prothoracic setae: 1-6, single; occasionally seta 1, 2 branched; 7, 3 branched. *Abdomen*. VIIIth segment: lateral comb patch of 27-36 coarsely fringed scales; seta 1, 1-3 branched; 2 and 4, single; 3, 5-9 branched, plumose; 5, 3-6 branched. Siphon tapering slightly towards apex; index 2·6-2·8; pecten of 15-21 close-set, dark thick spines with pale tips and one large and 3-5 smaller denticles at base; seta 1 arising about half length of siphon, 5-7 branched, plumose. Anal segment: a small separate elongate sclerotized plate at the lower proximal angle of saddle; saddle covering dorsal three-fifths of segment; seta 1 and 3, single; 2, 5-9 branched; 4 (ventral brush) of 16 tufts, 2 or 3 precratal. Anal papillae equal, pointed, about half length of saddle.

Biology

The most common breeding places of *calcariae* in Victoria are pits under uprooted trees, or occasional pits 2-3 feet deep, sometimes connected with crayfish tunnels or flooded rabbit burrows; usually such places are shaded by trees or shrubs. *Ae. calcariae* has been found mostly in low hill country or adjoining forested plains.

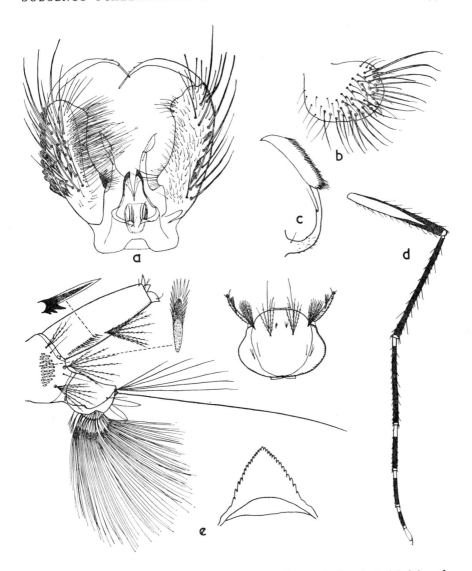

Fig. 36 *Aedes calcariae* Marks. *a-d*, adult: *a*, male terminalia; *b*, basal lobe of coxite; *c*, harpago; *d*, hind leg. *e*, larva: head, mentum and terminal segments.

It is a man-biting species but is not as persistent as *flavifrons;* it usually bites in cloudy weather or in the shade. Females continue biting activity after sunset, and as late as December (Lyonville).

DISTRIBUTION

Ae. calcariae is distributed on and south of the Central Highlands in Victoria; it is also recorded from the south-eastern part of South Australia.

Aedes (Ochlerotatus) purpuriventris Edwards

Aedes purpuriventris Edwards, 1926, *Bull. ent. Res.*, 17: 13. Dobrotworsky, 1960b, 85: 183.

The wing membrane has a dark blotch in the middle. The wings and legs are unmottled; the tarsi are unbanded. The venter is partly or wholly clothed with purple scales.

ADULT FEMALE

Vertex with narrow and upright forked, pale scales. Proboscis and palps purplish-black. Scutal integument dark brown. Scutum clothed with brown scales becoming paler around front margin and around bare area; scales sometimes pale brown or golden. Posterior pronotum (fig. 37 d) with broad scales below, narrow curved scales above. Scales on postspiracular area broad, flat. Wings dark scaled; dark blotch on wing membrane. Femora and tibiae black anteriorly, unmottled; hind femora dark anteriorly and posteriorly on apical quarter or half. Tarsi unbanded. Tergites purplish-black, unbanded, with small basal lateral patches of white scales. Venter may be purple or ochreous with more or less extensive mottling.

ADULT MALE

Palps slightly shorter than proboscis, purplish-brown scaled. *Terminalia* (fig. 37 a-c). Coxite about 3 times as long as broad with long, yellowish setae sternally, particularly on distal half. Apical lobe moderate with several short and a few longer setae. Basal lobe a rounded plate with numerous fine setae. Style about half length of coxite, curved, swollen for about half length; apical third slender. Harpago stout, curved with 2-3 long setae and numerous fine ones on basal half; appendage as long as harpago, fimbriated and with curved tip. Paraproct with single tooth. Phallosome simple. Lobes of IXth tergite with 3-5 strong setae.

LARVA (fig. 37 e, f)

Head, siphon and saddle light brown. Head about as long as broad. Antenna thin, curved, as long as or longer than head; seta 1 arising near mid length, 2-5 branched. Head setae: 4, 2-3 branched; 5, 4-6 branched; 6, 2-4 branched; 7, 6-11 branched; 8 and 9, 2 branched. Prothoracic setae: 1 and 3, 1-2 branched; 2, 4, 5 and 6, single; 7, 2 branched. *Abdomen.* VIIIth segment: lateral comb patch of 28-35 coarsely fringed scales; seta 1, 3-4 branched; 2, single or 2 branched; 3, 7-8 branched; 4, single; 5, 3-5 branched. Siphon short with index 2·3-3·0; pecten of 18-23 close-set dark spines distinctly paler towards tip and with one large stout, and 1-3 small, denticles at base; seta 1, 4-6 branched arising about mid length of siphon. Anal segment: saddle almost complete ring; small separate elongate sclerotized plate at lower proximal angle of saddle; setae 1 and 3, single; 2, 6-8 branched; 4 (ventral brush) of 16-17 tufts, 2-3 precratal. Anal papillae pointed, about half length of saddle; lower pair slightly shorter than upper.

BIOLOGY

Ae. purpuriventris is a spring species. It breeds in shaded ground pools

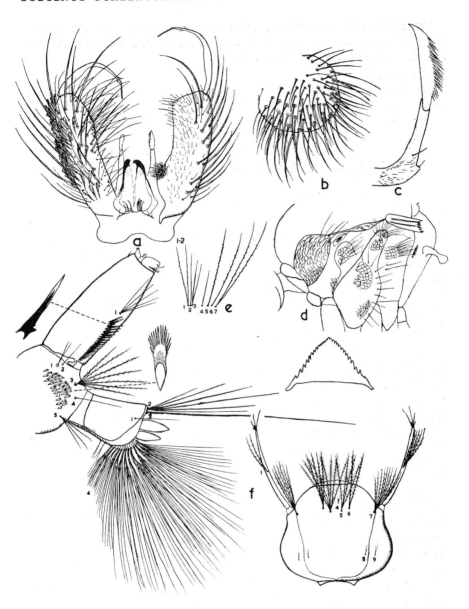

FIG. 37 *Aedes purpuriventris* Edwards. *a-d*, adult: *a*, male terminalia; *b*, basal lobe; *c*, harpago; *d*, side view of thorax. *e-f*, larva: *e*, prothoracic setae; *f*, head, mentum and terminal segments. (*a-c, e-f*, after Dobrotworsky.)

and shows a preference for pits 2-3 feet deep; larvae are also commonly found in pools under uprooted trees. In South Australia, larvae have been found breeding in flooded rabbit burrows. Adults have not been collected later than the end of January.

During the spring and early summer it is a day-biting mosquito, but on hot days in December or January it bites only after sunset. It does not bite man as readily as does *Ae. flavifrons*.

DISTRIBUTION

Ae. purpuriventris has been collected at a number of scattered localities in Victoria: Welshpool, Baxter, Maroondah, Bulla, Lyonville, Lorne, Cape Otway, Avoca, Maryvale. It also occurs in Tasmania.

Aedes (Ochlerotatus) clelandi (Taylor)

Culicada clelandi Taylor, 1914, *Trans. ent. Soc. Lond.*, 1913: 690. *Aedes clelandi* Edwards, 1924, *Bull. ent. Res.*, 14: 379. Dobrotworsky, 1960b, 85: 185.

The scutum is uniformly clothed with golden scales; all the femora are unmottled, hind pale on basal half to third anteriorly. The tarsi are unbanded, but on the hind ones there are some pale scales at the base of the segments. The tergites have narrow creamy basal bands.

ADULT FEMALE

Vertex clothed with curved pale scales and forked upright brown scales. Proboscis and palps black scaled. Scutal integument reddish-brown. Scutum clothed with golden scales with some scattered black ones. Posterior pronotum with narrow curved scales usually pale below, black above. Postspiracular area (fig. 38 c) with patch of narrow scales. Wings dark scaled. Legs (fig. 38 d) dark scaled; hind femora pale, dark anteriorly on apical half or third with dark dorsal line reaching almost to base. Tarsi unbanded, but hind ones with some pale scales at base of first 2-4 segments. Tergites (fig. 38 e) violet-black scaled, usually with basal creamy-white bands which may be reduced to a few pale scales; tergite VII may be mottled. Venter ochreous scaled; black scales may form central patch on sternites or may be scattered, giving a mottled appearance, or may be few.

ADULT MALE

Palps slightly longer than proboscis. Lower mesepimeral bristles usually absent. *Terminalia* (fig. 38 a, b). Coxite almost 4 times as long as broad with long golden mesially directed setae arising on apical half. Apical lobe prominent with short setae. Basal lobe a rounded plate with numerous fine setae. Style about half length of coxite, curved, swollen at about mid length; apical third slender. Harpago stout, curved, with prominent thumb near base; basal third with numerous fine setae and 3-4 larger ones; appendage almost as long as harpago, blade-like, fimbriated, tip blunt. Paraproct with single tooth. Lobes of IXth tergite with 3-5 strong setae.

LARVA (fig. 38 f, g)

Head, siphon and saddle light brown. Head about three-quarters as long as broad. Antennae short and stout, less than half length of head; seta 1 arising near mid length of antenna, 4-5 branched. Head setae: 4, 2-3 branched; 5, usually 3 branched; 6, single; 7, 6-8 branched; 8, 1-2 branched; 9, 2 branched. Mentum with 14-15 small lateral teeth. Prothoracic setae 1-6, single; 7, 2-3 branched. *Abdomen*. VIIIth segment:

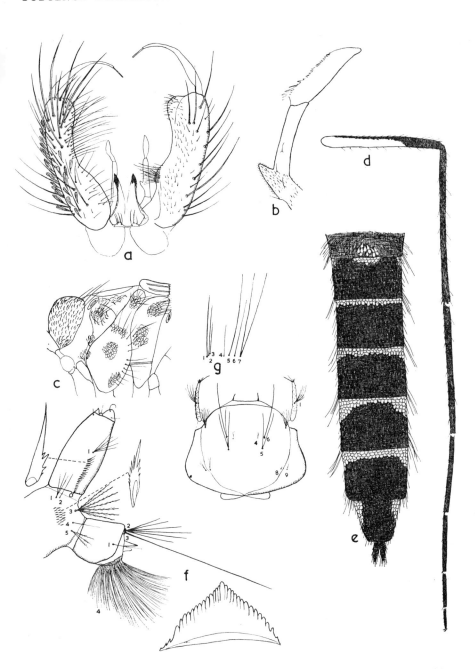

FIG. 38 *Aedes clelandi* (Taylor). *a-e*, adult: *a*, male terminalia; *b*, harpago; *c*, side view of thorax; *d*, hind leg; *e*, female abdomen, dorsal view. *f-g*, larva: *f*, head, mentum and terminal segments; *g*, prothoracic setae. (*a*, *b*, *f*, *g*, after Dobrotworsky.)

lateral comb of 19-21 scales fringed at base and with central spine twice as long as nearest lateral one; seta 1, 3 branched; 2 and 4, single; 3, 7-8 branched; 4, 3-4 branched. Siphon short, with index 2·0-2·5; pecten of 19-21 spines, each spine usually with one large, stout denticle and 1-3 small ones; seta 1, 5 branched arising about mid length of siphon. Anal segment: saddle complete ring; separate small, elongate sclerotized plate ventrolaterally near saddle; seta 1 and 3, single; 2, 6-7 branched; 4 (ventral brush) of 16 tufts. Anal papillae short, pointed, upper pair less than half length of saddle, lower pair shorter.

BIOLOGY

Ae. clelandi is a spring species. Larvae have not been collected in the field in Victoria but presumably the breeding sites are similar to those recorded elsewhere. In Western Australia *clelandi* breeds in fresh water pools with or without vegetation (Britten, 1955); in South Australia larvae have been found with those of *Ae. calcariae* in flooded rabbit burrows where the water may be 2 feet below ground level and heavily contaminated with rabbit faeces (Marks, 1957). Adults have been collected as early as July (Britten, 1955) but in Victoria they are most common during September-November. On Phillip Island they occur in numbers in the tea-tree scrub adjoining the shore. It is a day-biting species which attacks man.

DISTRIBUTION

Ae. clelandi has been collected at only two Victorian localities: Phillip Island and Lower Tarwin. The species occurs in Tasmania, Flinders Island, South Australia and Western Australia.

PERKINSI SECTION

Male terminalia. Apical lobe of coxite prominent, with fine hairs; basal lobe prominent with a short stout seta near base and a row of long recurved or straight setae at apex, the most tergal of which may be thickened; several rows of long mesially directed setae arising sternally. Harpago without thumb; appendage broad, blade-like.

Aedes (Ochlerotatus) perkinsi Marks

Aedes perkinsi Marks, 1949. *Pap. Dep. Biol. Univ. Qd,* 2 (11): 32.

Similar to *Ae. rubrithorax* and *Ae. procax* but readily distinguished from the former by having a pointed abdomen and from the latter by having black scaled proboscis and tergites with either triangular patches or laterally constricted basal bands. The scutum clothed with rather large golden scales. The tibiae are unmottled and the tarsi have white basal bands.

ADULT FEMALE

Vertex with narrow curved creamy scales; upright forked scales golden. Proboscis and palps black scaled. Scutal integument reddish-brown. Scutum clothed with rather large narrow curved golden scales with some fine bronze scales; scales around bare area paler. Posterior pronotum with

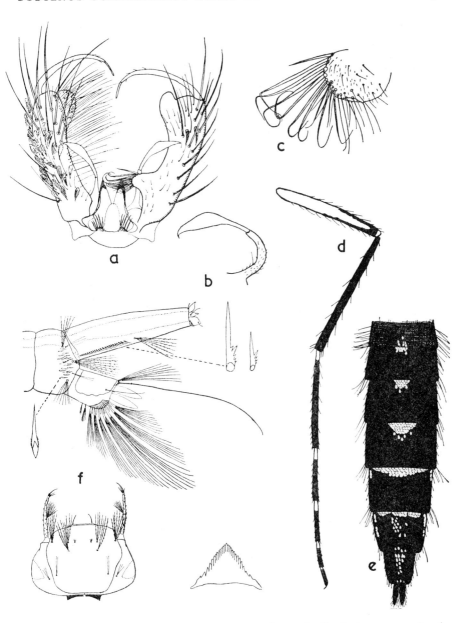

Fig. 39 *Aedes perkinsi* Marks. *a-e*, adult: *a*, male terminalia; *b*, harpago; *c*, basal lobe; *d*, hind leg; *e*, abdomen. *f*, larva: head, mentum and terminal segment (larva after Marks).

narrow curved black scales in middle, light-golden scales above and a patch of broad white scales below. 2-3 lower mesepimeral bristles. Wings dark scaled. Fore and mid femora mottled anteriorly; hind femur (fig. 39 d) pale on basal two-thirds with dorsal mottling and with apical third dark anteriorly and posteriorly. Knee spots golden. Tibiae dark dorsally, pale beneath. All tarsal segments with basal white bands. Tergites (fig. 39 e) black scaled with basal triangular patches of white scales medially and white lateral patches. Sternites white scaled with more or less conspicuous median and apical lateral black patches.

ADULT MALE

Palps about equal in length to proboscis with labella. *Terminalia* (fig. 39 a-c). Coxite black scaled, about 4½ times as long as broad with about 30 long setae along inner sternal edge. Apical lobe well developed with some fine setae. Basal lobe rounded, with row of about 6 long, curved setae and single short stout seta distally. Style three-fifths length of coxite, slender, slightly curved, with 2-3 short fine preapical setae; terminal appendage straight, about one-sixth length of style. Harpago stout; appendage about 1½ times the length of harpago, widening gradually then tapering off a short distance from tip. Paraproct with single tooth. Lobes of IXth tergite with 3-8 setae.

LARVA (fig. 39 f)

Head, siphon and saddle light brown. Head about two-thirds as long as broad. Antenna about half length of head; seta 1, 6-9 branched. Head setae: 4, small, 4-6 branched; 5, 3-5 branched; 6, 3-4 branched; 7, 5-8 branched; 8 and 9, 2-3 branched. Mentum with small median tooth and 11-13 lateral teeth. *Abdomen.* VIIIth segment: lateral comb patch of 20-23 fairly large pointed spines; seta 1, 5-6 branched; 2 and 4, 2 branched; 3, 9-10 branched; 5, 6-8 branched. Siphon slightly tapering, index about 3·7; pecten of 25-30 spines, each with 4-5 denticles basally; seta 1 arising slightly beyond mid length of siphon, 5-7 branched. Anal segment: saddle covering dorsal two-thirds of segment; surface with very fine spicules; setae 1 and 3, single; seta 2, 7-10 branched; 4 (ventral brush) of 16-18 tufts, 1 or 2 precratal. Anal papillae equal, lanceolate, slightly shorter than saddle.

BIOLOGY AND DISTRIBUTION

Larvae of *Ae. perkinsi* have not been collected in Victoria. In Queensland they have been found in peaty swamp pools in open heath country. It is a day-biting mosquito which attacks man. In Victoria adults have been collected in East Gippsland at Cabbage Tree Creek (December-February), Reedy Creek (April) and Genoa (February). It has also been recorded from South Queensland.

Aedes (Ochlerotatus) luteifemur Edwards

Aedes luteifemur Edwards, 1926, *Bull. ent. Res.,* 17: 112.
Dobrotworsky, 1960a, 85: 54.

The costal vein and vein R_1 are pale scaled towards the apices. The hind femora are ochreous toward the apex; the tarsi have a few pale scales at the base of the basal segments of the hind legs. The tergites are unbanded but are extensively mottled with ochreous scales toward the apex of the abdomen.

ADULT FEMALE

Vertex clothed with curved narrow creamy scales and upright forked scales which may be creamy or dark. Proboscis and palps more or less mottled. Scutal integument brown. Scutum clothed with golden scales, with some admixture of black ones. Posterior pronotum with small patch of elongate white scales below, black scales or mixture of elongate dark, and narrow curved, pale scales in middle, and narrow curved pale scales above. Wing scales (fig. 40 d) dark brown, except those towards apex of C and R_1 which are pale. Fore and mid femora dark anteriorly and mottled with pale scales; hind femur pale on basal half, becoming ochreous towards apex and with admixture of black scales in this area, particularly on dorsal side. All tibiae dark scaled with admixture of pale scales. Tarsi dark, unbanded, but with a few pale scales at bases of second and third segments of hind legs. Tergites purplish-black scaled, unbanded but with admixture of ochreous scales, increasing in number from tergite IV or V towards apex of abdomen. Venter white scaled with admixture of ochreous scales in middle and in apical angles of sternites; sternite VIII usually ochreous

ADULT MALE

Palps longer than proboscis with labella. *Terminalia* (fig. 40 a-c). Coxite about 3½ times as long as broad, with numerous long yellowish setae along inner sternal edge. Apical lobe prominent; basal lobe shelf-like, having on its margin one short stout seta and a row of about 13 long setae, about 7 of which are recurved apically. Style long, curved. Harpago stout with a few short setae at base; appendage long, blade-like, widening rapidly near middle and tapering towards tip. Paraproct with single tooth. Lobes of IXth tergite with 5-7 stout setae.

LARVA (fig. 40 f-i)

Head, siphon and saddle light brown. Head about two-thirds as long as broad. Antenna slightly more than half length of head, with scattered spicules; seta 1 arising about mid length, 7-9 branched. Head setae: 4, tiny, 5-6 branched; 5, 3-5 branched; 6, 2-4 branched; 7, 6-12 branched; 8, single or 3 branched; 9, 2-3 branched. Mentum with 13-14 lateral teeth on each side. Prothoracic setae: 1, 1-2 branched; 2 and 6, single; 3, 2-3 branched; 4 and 5, 2 branched; 7, 3 branched. *Abdomen*. VIIIth segment: lateral comb patch of 20-23 fringed scales, central tooth about twice as long as nearest lateral one; seta 1, 5-8 branched; 2, single, may be 2 branched; 3, 8-11 branched; 4, single; 5, 6-9 branched. Siphon tapering towards apex; index 2·9-3·4, mean 3·1; pecten of 21-24 spines each with 3-4 denticles at base, the distal one usually the largest; seta 1 arising about half-way along siphon, 6-9 branched, slightly plumose. Anal segment: saddle covering dorsal three-quarters of segment; surface with short arched rows of fine spines; a small separate sclerotized plate near lower proximal angle of saddle; seta 1, single; 2, 5-7 branched; 3, single; 4 (ventral brush) of 15-16 tufts, usually one precratal, but may be two. Ventral pair of anal papillae slightly shorter than dorsal pair, all pointed, less than half length of saddle.

BIOLOGY

Ae. luteifemur, which is one of the most common spring species, breeds

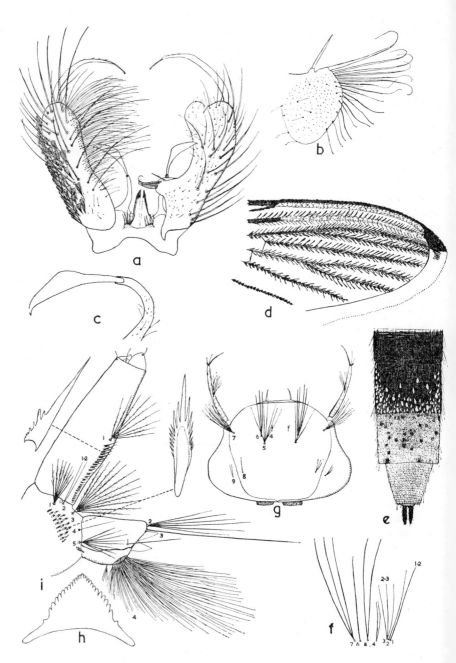

Fig. 40 *Aedes luteifemur* Edwards. *a-e*, adult: *a*, male terminalia; *b*, basal lobe of coxite; *c*, harpago; *d*, portion of wing; *e*, terminal segments of female abdomen, dorsal view. *f-i*, larva: *f*, prothoracic setae; *g*, head; *h*, mentum; *i*, terminal segments. (*a-c*, *f*, *g*, after Dobrotworsky.)

in temporary rain-filled pools more or less exposed to the sun. The pools, which vary in depth from a few inches to 2-3 feet, often have decayed leaves on the bottom and may have some vegetation. The breeding places are usually confined to sparsely wooded areas; in more heavily timbered country this species is confined to cleared patches where it breeds in pits left when trees have been uprooted. The eggs hatch immediately after the breeding places become filled by rain water. The adults usually appear in September and have been collected until early March. Apparently a second summer generation may develop when breeding places are filled by summer rains.

It is a day-biting mosquito attacking man, rabbits and domestic birds. Its biting activities are mainly restricted to the vicinity of the breeding places.

DISTRIBUTION

In Victoria *Ae. luteifemur* is widely distributed on and south of the Central Highlands. It is confined to woodlands, mainly on plains, or low hills; it has not been collected at altitudes greater than 2,000 feet. The species also occurs in Tasmania and Flinders Island.

Aedes (Ochlerotatus) silvestris Dobrotworsky

Aedes waterhousei Dobrotworsky, 1960, *Proc. Linn. Soc. N.S.W.*, 85: 57. *Aedes silvestris* Dobrotworsky, 1960, *Proc. ent. Soc. Wash.*, 62: 248.

A rather large species with black unbanded tarsi. The scutum is clothed with narrow dark-bronze scales mesially. The tergites have white convex basal bands.

ADULT FEMALE

Vertex clothed with curved narrow, and upright forked, creamy scales. Large elongate patch of narrow curved bronze scales close to posterior margin of eyes. Proboscis and palps black scaled. Scutal integument dark brown. Scutum between dorsocentral bristles, on inner half of fossa and on lateral areas clothed with narrow, almost straight dark-bronze scales; larger pale scales on outer part of fossa, on margin of lateral areas and around prescutellar bare area. Posterior pronotum with a few elongate pale scales below, narrow and elongate dark scales in middle, and curved narrow pale scales above. Fore and mid femora dark anteriorly, mottled with white scales. Hind femur (fig. 41 d) pale on basal three-quarters, dark scaled apically with some mottling. All tibiae dark scaled with a few pale scales. Knee spots conspicuous. Tarsi dark, first segment of tarsi with paler scales basally. Wing scales black. Tergites (fig. 41 e) with convex basal white bands. Sternite white scaled with elongate median black patch half as long as sternite and small apical lateral black patches.

ADULT MALE

Palps slightly longer than proboscis with labella. Tergites dark-brown scaled; basal white bands broken in middle but joining white lateral spots. *Terminalia* (fig. 41 a-c). Coxite about 3½ times as long as broad, with numerous long mesially directed golden setae arising sternally along the inner edge. Apical lobe prominent with several fine setae. Basal lobe shelf-like, having on its margin one short stout seta and row of 11-13 long setae, 6-8 of which are recurved apically. Style slender, curved. Harpago with a few moderately strong setae; appendage blade-like, expanded beyond mid length and then tapering to curved tip. Paraproct with single tooth. Lobes of IXth tergite with 4-7 short, stout setae.

LARVA (fig. 41 f-i)

Head, siphon and saddle dark brown. Head about three-quarters as long as broad. Antenna about half length of head, spiculate; basal quarter light brown, distal three-quarters darker; seta 1, 6-8 branched. Head setae: 4, tiny, 3-5 branched; 5, usually 3 branched, rarely 4; 6, usually 2 branched, rarely 3; 7, 5-8 branched; 8, usually single, may be 2 branched; 9, 2 branched. Mentum with small central tooth and 11-12 lateral teeth on each side. Prothoracic setae: 1, 2 branched; 2, 4 and 6, single; 3, usually 3 branched, rarely 2 or 4; 5, 2-3 branched; 7, 3 branched. *Abdomen*. VIIIth segment: lateral comb patch of 20-29 spines, fringed at base; central spine 2-3 times longer than any other; seta 1, 4-6 branched; 2 and 4, single; 3, 7-11 branched; 5, 4-6 branched; setae 1, 3 and 4 plumose. Siphon almost cylindrical up to distal spine of pecten, then tapering towards apex, index 2·6-3·0, mean 2·8; pecten on basal half of siphon, of 21-30 spines, mean 25; spines at base with 4-5 teeth; seta 1, 7-9 branched. Anal segment: saddle covering dorsal three-quarters of segment; surface with short arched rows of fine spines; a small separate sclerotized plate lies near lower proximal angle of saddle; seta 1, single, about as long as saddle; 2, 4-7 branched; 3, single; 4 (ventral brush), of 18-19 tufts, one precratal. Anal papillae narrow, about as long as length of saddle.

BIOLOGY

Ae. silvestris is most common in sparsely wooded shallow valleys or plains but also occurs in hill country at altitudes up to 2,500 feet. The larvae are found in pools 1-3 feet deep with clear or cloudy water and partly exposed to sun; they do not occur in heavily shaded pools in dense forest. In most places *Ae. silvestris* has only one generation per year. Larvae hatch immediately after adequate autumn-winter rains, and usually complete their development in early spring. The early appearance of adults will depend on the season and may occur as early as August. September is the more usual time and the adults remain numerous until November. A second generation is possible when heavy summer rains fill breeding places. At Steiglitz fourth-stage larvae and pupae have been collected in pools in a creek bed late in March and in the same month biting adults have been observed at Armidale (N.S.W.) and in the Grampians (Victoria).

Ae. silvestris is a day-biting species which is very numerous near its breeding places. During hot dry weather it bites only after sunset.

DISTRIBUTION

The distribution of *Ae. silvestris* sometimes overlaps that of *Ae. nivalis*, but it does not occur at such high altitudes as the latter species; it is apparently absent from Barrington Tops, Mt Kosciusko and Bogong High

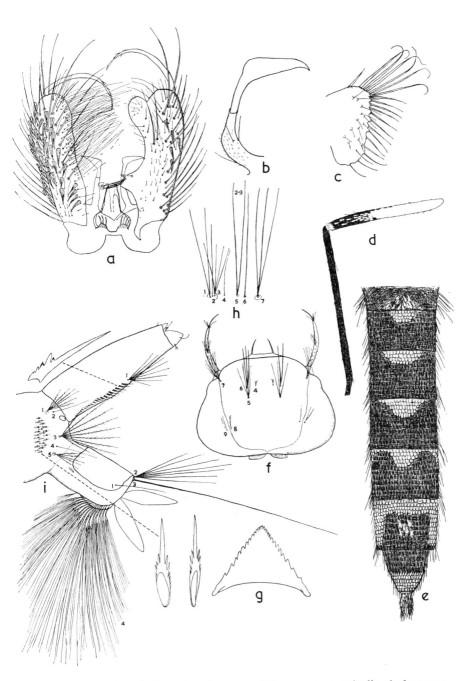

FIG. 41 *Aedes silvestris* Dobrotworsky. *a-e*, adult: *a*, male terminalia; *b*, harpago; *c*, basal lobe of coxite; *d*, hind femur and tibia; *e*, abdomen. *f-i*, larva: *f*, head; *g*, mentum; *h*, prothoracic setae; *i*, terminal segments. (*a-c*, *f-h*, after Dobrotworsky.)

Plains. On the other hand *silvestris* has a much wider distribution at lower altitudes. It occurs almost throughout Victoria but is absent from the Mallee and from treeless areas of northern parts of Victoria. *Ae. silvestris* is recorded from New South Wales and is common in the Mt Lofty Ranges in South Australia and in Tasmania.

Aedes (Ochlerotatus) nivalis Edwards

Cules australis Theobald, 1911 (*non* Erickson, 1842), Mon. Cul., 2: 91. *Menolepsis(?) tasmaniensis* Taylor, 1914, Proc. Linn. Soc. N.S.W., 39: 466. *Aedes nivalis* Edwards, 1926, Bul., ent. Res., 17: 112. Dobrotworsky, 1960a, 85: 60.

Similar to *Ae. silvestris* but readily recognized by two features: the scutal scales are golden and the bands on the tergites are almost straight.

ADULT FEMALE

Vertex clothed with narrow curved and upright forked scales, the latter becoming black towards the neck. Large elongate patch of narrow curved bronze scales close to posterior margin of eyes. Proboscis and palps black scaled; palps with a few pale scales. Scutal integument dark reddish-brown. Scutal scales narrow curved golden medially, broader and paler in front of wing roots and around prescutellar bare area; a few small black scales scattered on scutum. Fore and mid femora dark above, with some mottling. Hind femur (fig. 42 d) pale on basal two-thirds, black apically with a few scattered white scales. Tibiae black scaled with some scattered pale scales. Knee spots conspicuous. Tarsi black scaled with some pale scales on first segment. Wing black scaled. Tergites black scaled with almost straight, white basal bands. Sternites with central, and apical lateral, patches of black scales.

ADULT MALE

Palps as long as proboscis with labella, black scaled. Tergites black, hairy, with scales forming lateral patches only; no dark scales dorsally. *Terminalia* (fig. 42 a-c). Coxite about 4 times as long as broad, with numerous long golden setae arising sternally along inner edge. Apical lobe prominent, with several fine setae. Basal lobe shelf-like, having on margin one short stout seta, one long stout recurved seta and 4-6 finer setae, 3-4 of which have recurved tips. Style slender, curved. Harpago stout curved, with a few moderately long setae on basal half; appendage blade-like, expanding towards mid length and then tapering towards curved tip. Paraproct with single tooth. Lobes of IXth tergites with 3-6 short strong setae.

LARVA (fig. 42 f-l)

Head, siphon and saddle dark brown. Head about four-fifths as long as broad. Antenna about half length of head, spiculate; basal quarter light brown, distal three-quarters dark, almost black, but may be lighter towards tip; seta 1, 5-6 branched. Head setae: 4, tiny, 2-5 branched; 5, usually 3 branched, rarely 4 branched; 6, usually 3 branched, rarely 2 or 4 branched; 7, 5-9 branched; 8, single; 9, usually 2 branched, rarely single. Mentum with small central tooth and 11-13 lateral teeth on each side. Prothoracic

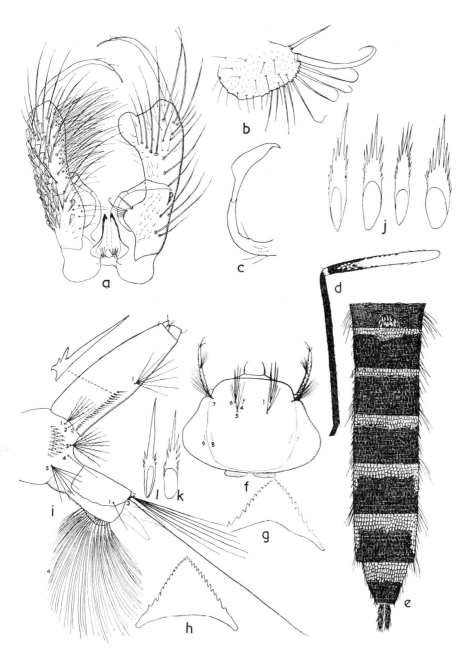

FIG. 42 *Aedes nivalis* Edwards. *a-e*, adult: *a*, male terminalia; *b*, basal lobe; *c*, harpago; *d*, hind femur and tibia; *e*, female abdomen, dorsal view. *f-l*, larva: *f*, head; *g*, mentum of larva from Marysville; *h*, mentum of larva from Lyonville; *i*, terminal segments; *j-l*, lateral comb teeth: *j*, from Bogong; *k*, from Marysville; *l*, from Lyonville. (*a-c, e-l*, after Dobrotworsky.)

setae: 1, 2 branched, rarely 3 branched; 2 and 6, single; 3, 3-4 branched, may be 2 or 5 branched on one side; 4, usually single, rarely 2 branched; 5, 2-3 branched; 7, 3 branched. *Abdomen.* VIIIth segment: lateral comb patch of 18-27 slender spines fringed at base; central spine 3-4 times as long as any other; seta 1, 4-7 branched; 2 and 4, single; 3, 8-11 branched; 5, 7-8 branched. Setae 1 and 3, plumose. Siphon almost cylindrical from base up to seta 1, then tapering towards apex; index 2·6-3·6, mean 3·0; pecten extending about half of length of siphon, with 20-26 spines, each spine usually with 3-4 small denticles and one larger one at base; seta 1, 6-8 branched. Anal segment: saddle covering dorsal half of segment; surface with fine spines arranged in short arched rows; small separate sclerotized plate near lower proximal angle of saddle; seta 1, single; 2, 5-8 branched; 3, single; 4 (ventral brush), of 16-17 tufts, one sometimes precratal. Anal papillae narrow, as long or longer than saddle.

Bogong Form

Specimens of *Ae. nivalis* from Bogong High Plains agree in general with *nivalis* from the type locality, but are smaller and darker. There are some differences in morphological traits of the larva: prothoracic seta 1 has 3-8 branches, with mean of 5 branches, whereas in larvae from other populations it has only 2-3 branches. The scales of the lateral comb are highly variable; they may have a single central tooth three times as long as the nearest lateral one or three long central teeth almost equal in size (fig. 42 j).

BIOLOGY

Ae. nivalis is confined mainly to elevated country and rarely descends to the plains. In Victoria *Ae. nivalis* breeds usually at altitudes higher than 2,000 feet, mainly in shallow pools with decayed leaves, but also in pools 2-3 feet deep. The water may be clear or cloudy. The pools are usually exposed to the sun or to dispersed light; only once have larvae been collected from a fully shaded pool. Although *nivalis* is associated with woodland it avoids dense forests with deep shade and is confined to patches of sparse bush or clearings. The Bogong form breeds on the High Plains at altitudes of 5,300-5,400 feet in pools 1-2 feet deep exposed to the sun. In New South Wales it has been recorded at Barrington Tops at 5,000 feet (I. M. Mackerras, 1927), Mt Kosciusko 6,000 feet (L. E. Cooling, 1924) and in the Armidale area (E. J. Waterhouse) at 3,300-4,800 feet.

It seems likely that in most localities *nivalis* has only one generation a year, because the breeding places remain dry during the summer months. However, the Bogong form may pass through two or more generations during the summer; it passes the winter in the larval stage under ice. Laboratory experiments have shown that larvae survive in water at 0°C.

At high altitudes, e.g. Marysville (3,000-3,200 feet), the larvae do not complete their development before the end of November and in the Mt Buller area (4,000 feet) larvae have been collected late in December. At lower altitudes (Lyonville) larval development is usually completed not later than the end of October.

DISTRIBUTION

In Victoria *Ae. nivalis* is distributed along the Central Highlands. It has been recorded from Bogong High Plains, Mt Hotham area, Mt Buffalo, Mt Buller area, Lake Mountain, Marysville, Trentham, Ballan, Lyonville and Mt Victory (Grampians). In New South Wales it has been found as far north as Ben Lomond.

Aedes (Ochlerotatus) camptorhynchus (Thomson)

Culex camptorhynchus Thomson, 1868, *Eugenie's Resta Dit.*, 443. *Culex labeculosus* Coquillett, 1906, *Ent. News,* 16: 116. *Culicelsa westralis* Strickland, 1911, *Entom.*, 44: 131. *Culicada inornata* Strickland, 1911, *Entom.*, 44: 201. *Culicada annulipes* Taylor, 1913, *Trans. ent. Soc. Lond.,* 693. *Culicada victoriensis* Taylor, 1914, *Proc. Linn. Soc. N.S.W.*, 39: 400. *Culicada nigra* Taylor, 1914, *Trans. ent. Soc.*, 688. *Aedes camptorhynchus* Edwards, 1924, *Bull. ent. Res.*, 14: 374. Dobrotworsky, 1960a, 85: 63.

The proboscis and legs are extensively mottled with pale scales; the tarsi are banded. The wing scales are dark. The tergal bands are convex.

Adult Female

Vertex clothed with narrow curved, and upright forked, pale scales; the latter may be darker laterally and towards neck. Proboscis and palps (fig. 43 d) extensively mottled, often almost completely pale scaled. Scutal integument dark brown. Scutum clothed with dark-golden or light-golden narrow curved scales with one or two white patches about mid length. Posterior pronotum with patch of moderately broad white scales below, narrow dark scales in middle and above. Femora, tibiae and first tarsal segments mottled (fig. 43 e). Tarsi with white basal bands; hind tarsal segment 5 also banded. Wing dark scaled. Tergites (fig. 43 f) black scaled with median patch on first and second segments and basal bands on segments 3-6; bands are usually convex but may be straight and narrow or triangular with apex sometimes reaching posterior margin of tergite; basal white lateral patches on all segments. Sternites white scaled with median and lateral apical patches of black scales; these patches may be reduced to a few black scales.

Adult Male

Proboscis usually with a few pale scales or may be entirely black scaled. Palps as long as proboscis with labella, dark scaled with white scales at base of segments 2, 4 and 5. *Terminalia* (fig. 43 a-c). Coxite about 4 times as long as broad, with numerous long, yellowish, mesially directed, sternal setae. Apical lobe large, with several moderately long setae. Basal lobe moderate, having on margin one short stout seta, one long, stout, recurved seta and 6-11 fine, short, straight setae. Style slender, curved. Harpago stout, curved; appendage narrow on basal two-thirds then, after sharply expanding, tapering towards tip. Paraproct with single tooth. Lobes of IXth tergite with 3-6 strong, short setae.

Larva (fig. 43 g, h)

Head light brown. Antennae less than half length of head, spiculate; seta 1, 3-6 branched. Head setae: 4, tiny, 5-7 branched; 5, 3-4 branched;

6 and 9, 2-3 branched; 7, 8-9 branched; 9, single or 2 branched. Mentum with small central tooth and 9-10 lateral teeth on each side. Prothoracic setae: 1, 2 branched; 2, 4, 5 and 6, single; 3 and 7, 2-4 branched. *Abdomen.* VIIIth segment: lateral comb patch of 24-33 fringed scales without long central tooth; seta 1, 4-5 branched; 2 and 4, single; 3, 7-8 branched; 5, 5-8 branched. Siphon with index 2·3-3·0; pecten of 22-24 spines each with 3-4 denticles at base; seta 1, 7-8 branched. Anal segment: saddle covering about dorsal half of segment with a small separate sclerotized

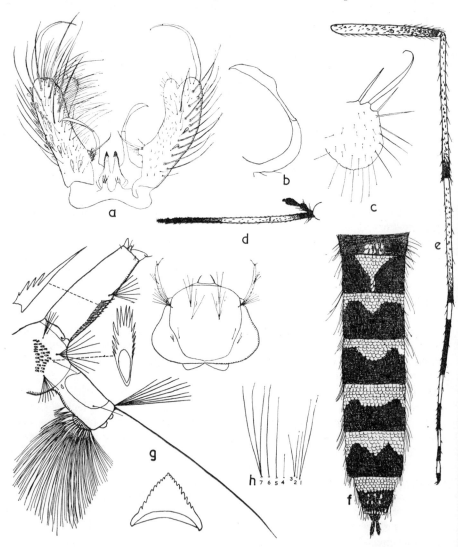

Fig. 43 *Aedes camptorhynchus* (Thomson). *a-f*, adult: *a*, male terminalia; *b*, harpago; *c*, basal lobe; *d*, proboscis and palp of female; *e*, hind leg; *f*, abdomen. *g-h*, larva: *g*, head, mentum and terminal segments; *h*, prothoracic setae. (*b, c, h*, after Dobrotworsky.)

plate near proximal angle; setae 1 and 3, single; 2, 6-7 branched; 4, of 18-19 tufts, usually 2 precratal. Anal papillae very small, globular.

BIOLOGY

Ae. camptorhynchus breeds mainly in brackish swamps in open country, but it can easily be carried by winds into areas where brackish water is absent. In such conditions *camptorhynchus* breeds, more or less successfully, in fresh water swamps; this has been recorded at Violet Town more than 80 miles from the nearest brackish swamps; Britten (1958) frequently collected larvae of this species in fresh water in Western Australia.

In Victoria *Ae. camptorhynchus* is a homodynamous species; all larval stages, pupae and the emergence of adults have been observed on many occasions during the winter.

It is a vicious mosquito, which attacks during the day and particularly after sunset. It has been recorded as biting man and domestic animals.

DISTRIBUTION

In Victoria *Ae. camptorhynchus* is not restricted to the coastal areas, but is permanently established inland wherever brackish waters occur. Thus in the Mildura area (about 230 miles from the coast) *camptorhynchus* is a very common species. Adults are particularly numerous in a woodland coastal belt about 20-30 miles wide; they are absent only in areas where mountains are close to the shore, leaving no space for brackish swamps; such a situation occurs in the Lorne-Apollo Bay area. This species is common in South Australia and Tasmania, and also occurs in New South Wales and Western Australia.

CUNABULANUS SECTION

Male terminalia. As in Perkinsi Section, but apical lobe is slightly less prominent and there are no rows of long mesially directed setae sternally on coxite.

Aedes cunabulanus Edw. has been reported for Victoria (Dobrotworsky, 1960) on the basis of a single, rather damaged specimen, a female, collected at Wilson's Promontory in 1953. All attempts to collect this species at that and other localities have failed and *Aedes cunabulanus* should be removed from the list of Victorian species until its presence is confirmed.

Aedes (Ochlerotatus) andersoni Edwards

Andersonia tasmaniensis, Strickland 1911, *Entom.,* 44: 250. *Menolepsis tasmaniensis* Taylor, 1914 (*non* Strickland), *Proc. Linn. Soc. N.S.W.,* 39: 466. *Aedes andersoni* Edwards, 1926, *Bull. ent. Res.,* 17: 112. Dobrotworsky, 1960a, 85: 68.

The typical form is easily recognized by the patch of broad flat white scales in front of each wing root; this is absent in the Grampians form. Both forms have a median stripe of small dark-bronze scales on the scutum.

ADULT FEMALE

Vertex clothed with narrow curved and upright forked, creamy scales.

Proboscis and palps black scaled. Scutal integument dark brown. Scutum (fig. 44 f) with median stripe of small narrow curved dark-bronze scales between the dorsocentral bristles broadening on posterior half and contrasting with a broad margin of creamy scales; a large patch of broad white scales in front of wing roots. Posterior pronotum with narrow curved white scales below, narrow curved and elongate dark, almost black, scales in middle and narrow creamy scales above. All femora (fig. 44 d) mottled. Fore and mid tibiae black anteriorly, with a few pale scales. Hind tibia black anteriorly with pale scales forming a streak. Tarsi dark, unbanded; first segment of hind tarsus mottled on basal three-quarters. Wing dark scaled. Tergites black scaled with white lateral patches. Sternites white scaled, with central, and apical lateral, patches of black scales; central patches sometimes absent.

ADULT MALE

Palps about as long as proboscis with labella. Tergites unbanded, with only lateral patches of white scales. *Terminalia* (fig. 44 a-c). Coxite 3½-4 times as long as broad with only a few moderately long setae directed mesially. Apical and basal lobes prominent; apical lobe with a few short seta, basal lobe shelf-like, having on its margin one short stout seta, one long stout seta with recurved tip and 10-14 long finer setae with recurved tips; remaining setae fine and straight. Style slender curved, tapering on distal third. Harpago stout, curved with slightly developed thumb near base of appendage; appendage lanceolate, as long as harpago. Lobes of IXth tergite with 3-7 short strong setae.

LARVA (fig. 44 g-j)

Head, siphon and saddle light brown. Head three-fifths as long as broad. Antenna almost half length of head, with scattered spicules; seta 1 arising at about mid length, 5-6 branched. Head setae: 4, tiny, 2-4 branched; 5, 3-5 branched, usually 4 branched; 6, usually 3 branched, rarely 2 or 4 branched; 7, 5-7 branched; 8, single; 9, 1-2 branched. Mentum with a median tooth and 11-12 lateral teeth on each side. Prothoracic setae: 1, 2 and 6, single; 3, 4 and 5, 2 branched; 7, 2-3 branched. *Abdomen.* VIIIth segment: lateral comb a patch of 24-33 fringed scales, central tooth twice as long as nearest lateral ones; seta 1, 5-6 branched; 2, 2-3 branched; 3, 9-10 branched; 4, 2 branched; 5, 7-8 branched; seta 3 plumose, all others simple. Siphon tapering towards apex; index 3·2-4·0, mean 3·7; pecten on basal half of siphon of 24-37 (mean 31) close-set, strong dark spines paler towards tip; seta 1 arising at three-fifths length from base, 4-9 branched, plumose. Anal segment: saddle covering dorsal three-quarters of segment with small elongate sclerotized plate near lower proximal angle; surface with coarse denticles on dorsodistal part; seta 1, single, about as long as saddle; 2, 9-11 branched; 3, single; 4 (ventral brush), of 18-19 tufts, 2 usually precratal. Anal papillae equal, pointed, less than half length of saddle.

Grampians Form

Specimens of *Ae. andersoni* from the Grampians, Victoria, agree in general with *andersoni* from Wilson's Promontory, but have no patch of broad

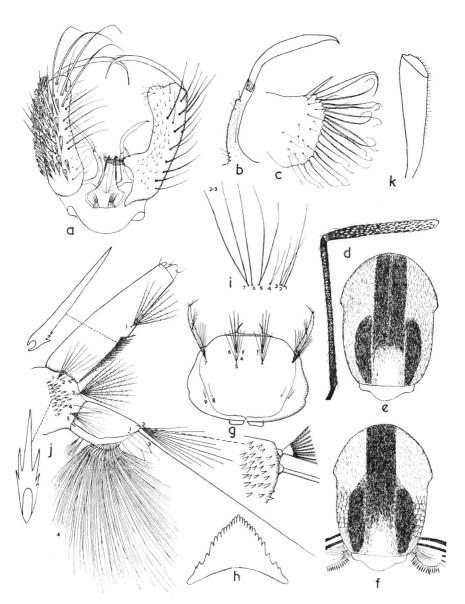

Fig. 44 *Aedes andersoni* Edwards. *a-f*, adult: *a*, male terminalia; *b*, harpago; *c*, basal lobe; *d*, hind femur and tibia; *e*, thorax of Grampians form; *f*, thorax of typical form. *g-j*, larva: *g*, head; *h*, mentum; *i*, prothoracic setae; *j*, terminal segments. (*a-c*, *g-k*, from Dobrotworsky.)

white scales in front of the wing roots (fig. 44 e). Posterior pronotum with white elongate scales below, narrow curved and elongate pale-goldish scales in the middle, and narrow curved pale scales above. Goldish scales along acrostichal bristles on scutum. Pale scales on tibiae arranged in longitudinal stripe. First segment of tarsi of all legs mottled with pale scales, particularly those of hind legs, which are almost entirely pale. Tergites 2 and 3 with trace of basal band.

BIOLOGY

The breeding places of *Ae. andersoni* are confined to sparsely wooded areas and particularly to tea-tree scrub. Larvae have been found in clean shallow roadside ditches and excavations as well as in a variety of shallow pools. At Wilson's Promontory it usually breeds in peat swamps in which the water is coloured dark by decayed moss. The adults appear as early as September and continue their biting activity through to autumn; biting females have been collected in March. It is a day-biting mosquito that attacks man, rabbits and birds.

DISTRIBUTION

The typical form of *Ae. andersoni* occurs in a coastal belt of Victoria some 20-30 miles wide (Wilson's Promontory, Tarwin Lower, Carpendeit, Cape Otway, Homerton, Gorae West, Timboon). It is also very common in Tasmania and Flinders Island. The Grampians form occurs as a pure population in three isolated localities in the Dividing Ranges: The Grampians and Bright, both in Victoria, and Kiandra in New South Wales. However, a mixed population of both forms and also intermediates (which have only a few broad white scales in front of the wing roots) occur in the Western District south of the 38th latitude.

Aedes (Ochlerotatus) continentalis Dobrotworsky

Aedes continentalis Dobrotworsky, 1960, *Proc. Lin.. Soc. N.S.W.*, 85: 71.

Rather closely resembles *Ae. cunabulanus* but is readily distinguished from it by the fact that the proboscis is mottled on its basal half.

ADULT FEMALE

Vertex clothed with narrow curved, and upright forked, creamy scales, the latter becoming black laterally and towards neck. Proboscis (fig. 45 d) mottled on basal three-quarters; palps mottled with creamy scales. Scutal integument brown. Scutum uniformly clothed with narrow curved golden scales. No bronze scales on fossa. Posterior pronotum with small patch of curved pale scales below, elongate blackish scales in middle, and narrow curved pale scales above. Scales on postspiracular area mostly narrow. All femora (fig. 45 e) mottled anteriorly and pale scaled posteriorly. Fore and mid tibiae mottled anteriorly; hind tibia with some mottling. Tarsi black scaled, unbanded, with first segment on all legs, and base of second segment of hind legs, mottled. Wings dark scaled. Tergites black scaled, usually with basal bands complete only on tergites IV and V; bands may be reduced to a few pale scales. Sternites white scaled, with median, and apical lateral, black patches; sternites III and IV may be mottled.

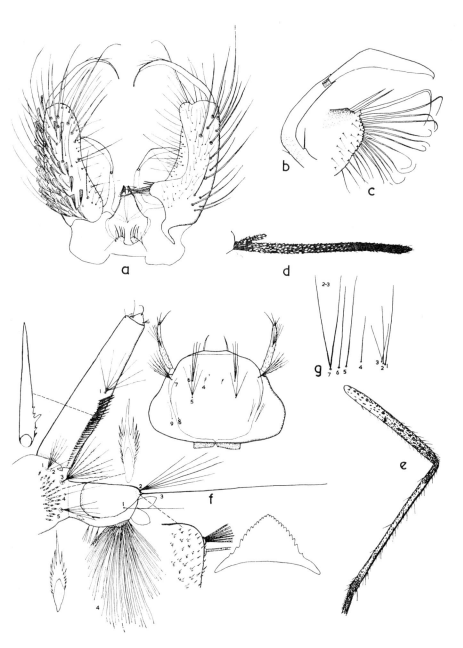

Fig. 45 *Aedes continentalis* Dobrotworsky. *a-e*, adult: *a*, male terminalia; *b*, harpago; *c*, basal lobe; *d*, female proboscis and palp; *e*, hind femur and tibia. *f-g*, larva: *f*, head, mentum and terminal segments; *g*, prothoracic setae. (*a-c*, *f*, *g*, after Dobrotworsky.)

ADULT MALE

Palps slightly longer than proboscis with labella, dark scaled, segments with creamy scales at base. *Terminalia* (fig. 45 a-c). Coxite about 3½ times as long as broad with only a few long setae directed mesially. Apical lobe prominent, with a few fine setae. Basal lobe shelf-like having on its margin 2 short spine-like setae and 12-14 long setae, most of them with recurved tips. Style curved, tapering apically. Harpago stout; appendage widening gradually towards half length and then tapering gradually towards tip. Paraproct with single tooth. Lobes of IXth tergite with 3-5 short setae.

LARVA (fig. 45 f, g)

Head, siphon and saddle light brown. Head about two-thirds as long as broad. Antenna almost half length of head; seta 1, arising at about mid length, 4-5 branched. Head setae: 4, small, 2-4 branched; 5, usually 3, may be 4, branched; 6, 3 branched; 7, 5-7 branched; 8 and 9, single or 2 branched. Mentum with small median tooth and 11-12 lateral teeth. Prothoracic setae: 1 and 2, single; 3, 2 branched; 4, 5 and 6, single; 7, usually 2, sometimes 3 branched. *Abdomen*. VIIIth segment: lateral comb patch of 24-32 fringed scales; central tooth twice as long as two nearest lateral teeth, or longer; seta 1, 5-6 branched, slightly fringed; 2 and 4, single; 3, 6-11 branched, fringed; 5, 4-6 branched. Siphon tapering towards apex, index 4·0-5·0, mean 4·5; pecten on basal half of siphon of 25-35 strong, closely set, dark spines with pale tips and 4-5 denticles at base; seta 1 arising slightly beyond mid length of siphon, 5-7 branched. Anal segment: saddle covering dorsal three-quarters of segment with small elongate sclerotized plate near lower proximal angle; surface with coarse irregular denticles on dorsodistal part; setae 1 and 3, single; seta 2, 5-7 branched; 4 (ventral brush), of 16-19 tufts, 2-4 precratal. Anal papillae equal, broad, about one-third length of saddle.

BIOLOGY

Ae. continentalis has been found breeding in rain-water pools and small swamps with grassy edges, exposed to the sun. Fourth-stage larvae have been collected as early as September. Adults have been collected from November to February but it is not a very common species. It is a day-biting mosquito which attacks man.

DISTRIBUTION

The distribution of *Ae. continentalis* in Victoria is similar to that of *Ae. andersoni;* it occurs in a coastal belt some 40 miles wide, but is absent from the Grampians. It also occurs in South Australia, Flinders Island and Tasmania.

STRICKLANDI SECTION

Male terminalia. Apical lobe of coxite prominent, with short broad setae; basal lobe scarcely developed, with six shortish setae; no rows of long mesially directed setae. Harpago with bristle but no distinct thumb; appendage short, broad, curved.

Aedes (Ochlerotatus) stricklandi (Edwards)

Grabhamia australia Strickland, 1911. *Entom.*, 44: 133.
Ochlerotatus stricklandi Edwards, 1912, *Ann. Mag. nat. Hist.*, 9: 523 (*nom. nov.*, for *australis* Strickland, *non* Erickson, 1842; *Ochlerotatus*). *Grabhamia flindersi* Taylor, 1914, *Trans. ent. Soc. Lond.*, 1913: 686. *Aedes stricklandi* Edwards, 1924, *Bull. ent. Res.*, 14: 377. Marks, 1963b, 2: 38.

Distinguished from all other species by the following combination of characters: the scutum is uniformly clothed with dark-bronze scales, contrasting with the white scaled pleura; the wings are mottled with broad white scales; the tarsi are unbanded; the tergites have rather few pale scales along apical border.

ADULT FEMALE

Vertex with broad band of curved brown scales behind eyes and creamy curved scales in middle; upright scales all dark. Proboscis with white mottling. Scutum clothed with dark-bronze narrow curved scales. Posterior pronotum with narrow dark-bronze scales and patch of broader elongate white scales below. Scutellum with white narrow curved scales. Pleuron (fig. 46 a) densely clothed with flat white scales. 1-2 lower mesepimeral bristles. Wings mottled with broad white scales. Legs dark; femora, tibiae and first tarsal segments mottled with white scales; some white scales at base of hind tarsal segments. Claws toothed. Tergites brown scaled with apical narrow white bands and basal, and apical lateral, white patches. Sternites white scaled with scattered dark scales.

ADULT MALE

Palps about as long as proboscis with labella, black scaled with white mottling. *Terminalia* (fig. 46 b, c). Coxites with prominent apical lobe bearing 8-10 short broad setae. No distinct basal lobe. Style slightly curved, about half length of coxite with 4 short fine preapical setae; appendage slender, about one-eighth length of style. Harpago stout, with 3 preapical setae; appendage a little longer than harpago, very broad, with short fimbriations along its sternal margin. Paraproct with a single tooth, and with minute denticles distally. Lobes of IXth tergite with 3-5 stout setae and 2-4 fine setae lateral to the lobe.

LARVA (fig. 46 d-h)

Head darkish brown, three-fifths as long as broad; fronto clypeus with 3 paler lenticular patches. Antenna about three-fifths length of head, spiculate; seta 1, 5-6 branched. Head setae: 4 and 7, 4-5 branched; 5 and 6, single; 8, 2-3 branched; 9, 3-4 branched. Mentum with 11-13 lateral teeth. Prothoracic setae; 1 and 2, single; 3, usually 2 branched; 4, 1-2 branched; 5 and 6, single; 7, 3 branched. *Abdomen*. VIIIth segment: lateral comb a row of 4 simple spines one-third length of saddle, arising from a sclerotization; seta 1, single; 2, 2-3 branched; 3, 6-8 branched; 4, 3-4 branched; 5, 4-5 branched; setae 1, 3 and 5 with sclerotized bases. Siphon tapering slightly on distal two-fifths with index $3 \cdot 3$; pecten extending over basal half, of 31-38 close-set spines, usually with 3-4 basal

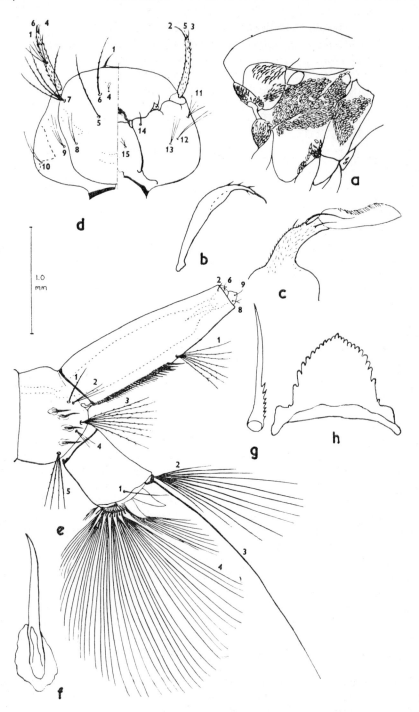

FIG. 46 *Aedes stricklandi* (Edwards). *a-c*, adult: *a*, side view of thorax; *b-c*, male terminalia: *b*, style; *c*, harpago. *d-h*, larva: *d*, head; *e*, terminal segments; *f*, lateral comb tooth; *g*, distal pecten tooth; *h*, mentum. (*a-g*, after Marks.)

denticles; seta 1 arising at three-fifths length of siphon, 4-7 branched, fringed. Anal segment: saddle a complete ring, surface spiculate; seta 1 single; 2, 10-15 branched; 4 (ventral brush) of 16 tufts, 1-2 precratal. Anal papillae subequal, upper about half length of saddle, lower shorter.

BIOLOGY

This is a day-biting species which attacks man.

DISTRIBUTION

Ae. stricklandi is a rare species in Victoria; it has been collected only at Lower Tarwin, Wilson's Promontory, Anglesea and Carpendeit. It occurs also in Western Australia, South Australia and Bass Strait islands.

Aedes (Ochlerotatus) spilotus Marks

Aedes spilotus Marks, 1963, *J. ent. Soc. Qd,* 2: 31-47.

A very large mosquito. The scutum is like that of *Ae. stricklandi* but this species is readily distinguished from others by the extensive mottling of wings, legs and tergites, with creamy scales.

ADULT FEMALE

Vertex clothed with narrow black scales behind eyes and pale narrow scales medially; upright forked scales black. Proboscis black scaled and with mottling of white scales. Palps black with white mottling. Scutal integument brown. Scutum clothed with small dark bronze scales with admixture of white scales; two small patches of white scales laterally near mid length, and one in front of wing base. Posterior pronotum with dorsal half clothed with narrow curved dark-bronze scales, ventral half with narrow brown, and white, scales above lanceolate white scales below. 2-4 lower mesepimeral bristles. Wing dark scaled with extensive mottling of larger creamy scales. Legs mottled with creamy scales; tarsi dark scaled, unbanded but basal segments mottled. Claws of fore and mid tarsi toothed; hind claws simple. Tergites dark scaled, unbanded but extensively mottled with creamy scales and with lateral basal and apical patches. Sternites pale scaled with small lateral apical black patches and some scattered black scales.

ADULT MALE

Palps as long as proboscis with labella or slightly longer, extensively mottled. *Terminalia* (fig. 47 b, c). Coxite with a prominent apical lobe bearing 4-5 stout setae. No distinct basal lobe. Tergally coxite bears numerous short fine setae, with a few longer ones at inner lower angle; a few longer setae tergally near apex, and numerous long setae laterally. Style curved, about half length of coxite with 2-3 preapical setae; appendage slender, about one-eighth length of coxite. Harpago stout, with 1-2 preapical setae; appendage broad, with rounded tip and with short fimbriations along its sternal margin. Paraproct with a single tooth and 5-6 small setae mesially. Lobes of IXth tergite with 3-5 stout setae and with 2-4 fine setae laterally to them.

LARVA (fig. 47 d-g)

Head two-thirds as long as broad, light brown; fronto clypeus with 3

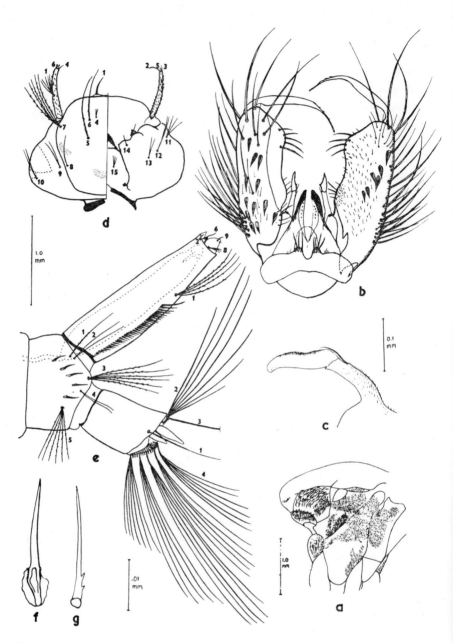

Fig. 47 *Aedes spilotus* Marks. *a-c*, adult: *a*, side view of thorax; *b*, male terminalia; *c*, harpago. *d-g*, larva: *d*, head; *e*, terminal segments; *f*, lateral comb tooth; *g*, distal pecten tooth. (*a, c-g*, after Marks.)

darker lenticular patches. Antenna slightly more than half length of head, spiculated; seta 1, 2-3 branched, arising slightly beyond mid length. Head setae: 4, 2-4 branched; 4 and 6, single; 7, 4-6 branched; 8, single; 9, single or 2 branched. Prothoracic setae; 1, 2, 4 and 5, single; 3 and 6, 2 branched; 7, 3 branched. *Abdomen.* VIIIth segment: lateral comb of 5 strong spines; seta 1, single; 2 and 4, 2 branched; 3, 6-8 branched. Siphon long, with index about 4·0; pecten of 32-43 spines, each with 3-6 almost equal teeth at base; seta 1 arising beyond mid length, 4-5 branched. Anal segment: saddle complete ring; setae 1 and 3, single; 2, 6-10 branched; 4 (ventral brush) of 15 tufts. Anal papillae, broad, pointed, almost one-third length of saddle.

BIOLOGY

Ae. spilotus breeds in a variety of ground pools. They are usually shallow and grassy and may be fully exposed to the sun or partly shaded; the water may be clear or cloudy. It is a day-biting species which attacks man.

DISTRIBUTION

It is widely distributed in the Western District of Victoria. It was collected in the following localities: Dunrobin, Derholm, Apsley, Maryvale, Cherrypool, Glenisla, Cavendish, Victoria Valley, Avoca and Maryborough. It is also recorded in South Australia.

Subgenus FINLAYA Theobald

Finlaya Theobald, 1903, *Mon. Cul.,* 3: 281. For synonyms
see Edwards (1932).

This is a very large subgenus with about 170 species, but only 10 of these occur in Victoria. The subgenus has a world-wide distribution but it attains its greatest development in the Oriental Region. Most species breed in small collections of water above ground such as in tree-holes, leaf bases of plants, artificial containers, etc.; a few including the common Victorian species breed in ground and rock pools.

Characters of the Subgenus

ADULT

Male palps usually slightly shorter than proboscis, rarely equalling it. Female palps usually one-fifth or one-eighth as long as proboscis, but sometimes two-thirds as long. Proboscis usually longer than front femora. Antennae of male with plume hairs directed mostly dorsally and ventrally. No lower mesepimeral bristles. Fore and mid tarsal claws toothed; hind claws simple in both sexes. VIIIth abdominal segment of female rather large, only slightly retractile and somewhat compressed laterally; sternite large and prominent; cerci always short. *Male terminalia:* Coxite without apical lobe; basal lobe feebly developed or absent. Harpago well developed, appendage long and flattened, rarely reduced to a bristle.

LARVA

Antennae smooth or spiculate. Head seta 6 in front of seta 5. Siphon usually short; seta 1 usually near middle. Lateral comb of a large triangular patch of scales or a single regular row of scales. Ventral brush distinct. On the basis of ornamentation, Edwards (1932) divided the subgenus into eight groups of species but Knight and Marks (1952) have since reduced the number to seven. In Victoria, three of these groups (A, E and F) are represented and there is, in addition, one species, *Ae. subauridorsum,* of undetermined position.

GROUP A (KOCHI-GROUP)

The Kochi-Group has apparently had two centres of dispersal, one Malayan, and the other Papuan (Marks, 1961). Of the 32 described species, 29 are confined to the tropics of the Australian region and the eastern part of the Oriental Region. Two others, essentially tropical species, have extended their range southwards, one to southern Queensland, and one to central New South Wales. In view of this pattern of distribution it is surprising that the remaining species, *Ae. dobrotworskyi,* should be confined to southern Victoria.

Characters of the Group

ADULT

Wings profusely spotted with areas of pale and dark scales. Scutum with a more or less definite variegated pattern of dark and pale scales. Femora and tibiae spotted and ringed with pale scales for nearly their whole length. Hind tarsi variously banded or marked with pale scales.

LARVA

Head setae 4 and 6 arise in front of the level of antennal bases; 7, usually on level behind that of 4 and 6; 5, posterior to that of 7. Lateral comb scales in a triangular patch.

Aedes (Finlaya) dobrotworskyi Marks

Aedes dobrotworskyi Marks, 1958, *Proc. Roy. Soc. Qd,* 69: 65.

A very small, rather dark, mosquito extensively marked with white. The proboscis has a white band near the middle and a white tip. The femora, tibiae and tarsi have white bands; the fifth tarsal segment of all legs is entirely white.

ADULT FEMALE

Vertex with flat broad scales with some narrow ones in midline; upright scales dark. Palps about a fifth or quarter length of proboscis, with large white patch at tip. Proboscis (fig. 48 g) dark with median white band for about one-fifth of its length, and small apical white band. Integument dark. Scutal scales narrow curved, mainly golden; a large patch of dark scales on fossa, a smaller patch in front of wing root, and in centre of scutum; several small patches of white scales laterally; white scales around prescutellar area are broad. Scutellum with flat, white and creamy scales. Posterior pronotum with broad white scales and some dark ones above. Broad white scales form a more or less continuous band across anterior pronotum, subspiracular area, upper sternopleuron and upper half of mesepimeron. 1-4 postspiracular bristles. Wings: C usually with basal, sector, subcostal, and apical pale areas; R_1 usually without prehumeral pale area; base of Cu usually dark; base of An sometimes dark. Knob of haltere black scaled. Leg (fig. 48 h) femora black scaled with 5-7 pale bands or patches and pale knee spot; tuft of outstanding dark scales ventrally at apex. Tibiae with 5-9 white bands or patches. Tarsi dark; tarsal segment 1 with basal, median and apical white bands; segment 2 of fore and mid tarsi white on apical half or quarter; 3 and 4 dark; 5 white. Segments 2 and 3 of hind tarsi white on apical half or third; 4 dark; 5 white. Tergites II-VIII usually with dorsal pale patches and white line extending from base along lateral border; VIII, with large basal white patch. Sternites dark scaled, with basal lateral white patches and creamy and white scales mesially; sternites V-VII usually with outstanding dark scales apically.

ADULT MALE

Palps slightly longer than proboscis with labella, dark scaled with 3 white bands on shaft; basal white patch on penultimate segment; basal white band and apical patch on terminal segment. *Terminalia* (fig. 48 a). Coxite dark scaled, 3 times as long as broad. Mesially on sternal aspect near middle is a row of 4-6 slender pointed scales; just tergal to scale tuft is dense patch of fine straight setae. Style about half length of coxite, moderately expanded on basal two-thirds, narrowing apically; appendage slender, curved, about half length of style. Harpago slender with short preapical seta; appendage expanding slightly near middle, slender, with fine curved tip. Paraproct with single tooth. Phallosome simple, with expanded and flattened apex. Lobes of IXth tergite with 1-3 setae.

LARVA (fig. 48 b-f)

Head, siphon and saddle light brown; thorax and abdomen with stellate setae. Head about as long as broad. Antenna short, straight, smooth; seta 1 arising about three-quarters from base, 2 branched, simple. Head setae: 1, 3-4 branched; 4, 5-10 branched; 5, single; 6, 3-4 branched; 7, 10-15 branched; 8, 2 branched; 9, 2-4 branched; 4, 6 and 7, arising in a line. Mentum with 9-10 pointed lateral teeth on each side. Prothoracic setae: 1, 13-20 branched, stellate; 2, single; 3, 17-24 branched. *Abdomen.* VIIIth segment: lateral comb of about 60-75 teeth arranged in 4 rows; setae 1 and 5, 16-24 branched, frayed, stellate; 2 and 4, single, simple; 3, 2-5 branched, frayed. Siphon pilose, with index 2·9-3·3: pecten of 7-9 broad fine-fringed spines; seta 1, 2-4 branched, arising near mid length of siphon. Anal segment: saddle covering more than half of segment, finely pilose, with 7-8 long slender spines on its distal margin above seta 1, and 3-6 below it; seta 1, long, 4-8 branched; 2, 2-4 branched; 3, single; 4, of 10 tufts; no grid. Anal papillae unequal, upper longer, and about 1-1½ times as long as saddle.

BIOLOGY

Larvae were collected from leaf bases of sword grass (*Gahnia* sp.). Adults have been collected from November until March in large patches of sword grass in swampy bush or along rivers. This is a day-biting mosquito which attacks man, usually on the face.

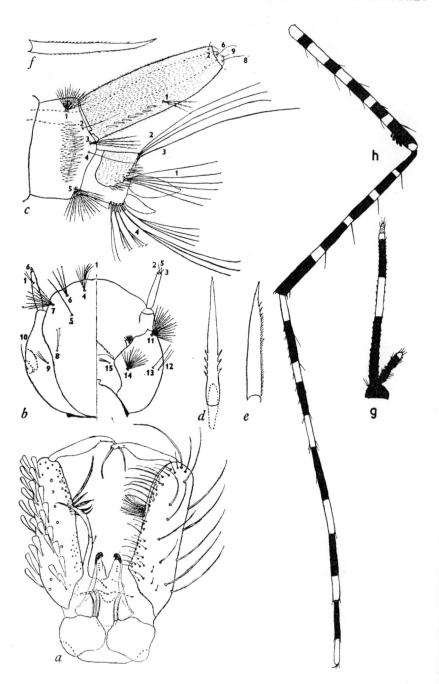

Fig. 48 *Aedes dobrotworskyi* Marks. *a, g-h*, adult: *a*, male terminalia; *g*, female proboscis and palp; *h*, hind leg. *b-f*, larva: *b*, head; *c*, terminal segments; *d*, tooth from distal row of lateral comb; *e*, pecten tooth; *f*, saddle spine. (*a* and *b-f*, after Marks.)

DISTRIBUTION

Ae. dobrotworskyi has been found in southern Victoria from Apollo Bay to Genoa. A single female was collected near Granya about 25 miles east of Wodonga; this is the only known locality north of the Central Highlands.

GROUP E (MEDIOVITTATUS-GROUP)

Thirteen species of this group are distributed in the Australian, Oriental and Neotropical (one species) Regions, three being present in Victoria.

Characters of the Group

Wings dark scaled. Scutum with pattern of white or yellow narrow longitudinal lines. Femora and tibiae with narrow line of white scales. Hind tarsi with 4 or 5 basal bands. All species breed in tree holes, bamboos, rock pools and artificial containers.

Aedes (Finlaya) notoscriptus (Skuse)

Culex notoscriptus Skuse, 1889, *Proc. Linn. Soc. N.S.W.*, 3: 1738. *Stegomyia notoscripta* Theobald, 1901, *Mon. Cul.*, 1: 286. *Aedes notoscriptus* (Skuse) Edwards, 1924, *Bull. ent. Res.*, 14: 382.

A small black mosquito conspicuously marked with narrow silvery-white scales on the scutum forming a lyre-shaped pattern, a long median line and short lateral lines. The proboscis and tarsi are banded.

ADULT FEMALE

Vertex with broad black scales, black upright scales and narrow silvery scales medially and around margins of eyes. Torus and tip of palps with patch of silvery scales. Proboscis (fig. 49 c) with white ring near middle. Scutum (fig. 49 d) clothed with dark-bronze, almost black, narrow scales with thin lines of narrow silver scales: a central line with fork around bare area, lateral lines forming a lyre-shaped pattern, a short line anteriorly between them and a second short line above wing roots. Scutellum with broad silvery scales. Posterior pronotum with broad black scales and patch of silvery scales below. Pleuron with several rounded patches of broad silvery scales. Wings dark scaled. Legs (fig. 49 e) black scaled; femora and tibiae with silvery line in front; fore femora white basally; knee spot white. Tarsi banded; last segment of hind tarsus white above. Tergites with basal laterally constricted creamy bands, sometimes greatly reduced; bands separated from lateral silvery spots. Venter pale scaled basally, gradually becoming dark scaled towards apex.

ADULT MALE

Palps about as long as proboscis with labella; penultimate and terminal segments short, each with white patch basally. Tergites may be unbanded; sternites black scaled with lateral silvery patches and some admixture of pale scales medially. *Terminalia* (fig. 49 a, b). Coxite cylindrical; basal and apical lobes absent; mesially numerous short setae; distal third of sternal margin with row of 20-30 long slender setae with flattened lanceolate tips.

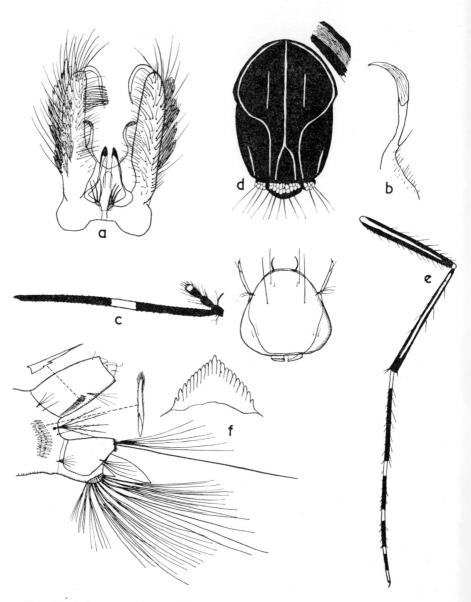

Fig. 49 *Aedes notoscriptus* (Skuse). *a-e*, adult: *a*, male terminalia; *b*, harpago; *c*, proboscis and palp of female; *d*, thorax; *e*, hind leg. *f*, larva: head, mentum and terminal segments.

Style short, curved; terminal appendage almost one-third length of style. Harpago curved, slender; appendage broad at base, slender on apical third and with pointed tip. Paraproct with single tooth. Lobes of IXth tergite with 3-4 strong, long setae.

Larva (fig. 49 f)

Head light brown, saddle and siphon dark brown. Antenna short, straight, smooth; seta 1 well beyond mid length, single. Head setae: 4, small tuft; 5 and 6, single; 7, 3-4 branched; 8, single or 2 branched; 9, 2-3 branched. Prothoracic setae: 1, 4 and 7, 2 branched; 2, 5 and 6, single; 3, 3 branched. *Abdomen*. VIIIth segment: lateral comb of 19-24 fringed spines arranged in 2 or 3 rows; seta 1, 2-4 branched; 2 and 4, single; 3 and 5, 4-5 branched. Siphon short, index about 2; pecten of 12-13 spines; seta 1, 3-4 branched, Anal segment: saddle covering about three-quarters of dorsal part of segment, with apical spines; seta 1, 2-3 branched; 2, 3 branched; 3, single; 4 of 12 tufts, each tuft 2-6 branched. Anal papillae pointed, unequal; upper pair about as long as saddle; lower pair half as long.

Biology

Larvae are found in tree holes or water tanks, occasionally in rock and ground pools. Overwintering occurs in the larval stage. *Ae. notoscriptus* is a day-biting species which attacks man and marsupials.

Distribution

This species is widespread in Australia and also occurs in New Guinea, New Caledonia, New Britain and New Zealand.

Aedes (Finlaya) mallochi Taylor

Aedes mallochi Taylor, 1944, *Proc. Linn. Soc. N.S.W.*, 69: 121. *Aedes pulcherrimus* (Taylor) Edwards, 1924, *Bull. ent. Res.*, 14: 382 (*non Mimeteomyia pulcherrima* Taylor, 1919, *Proc. Linn. Soc. N.S.W.*, 43: 830). Marks, 1955, 1: 13.

Similar to *Ae. notoscriptus* but is easily distinguished from it by the unbanded proboscis and by the fact that the scutal pattern is formed of broad white scales.

Adult Female

Similar to *Ae. notoscriptus* but may be recognized easily by the following traits. Proboscis (fig. 50 c) without white band. Scutum (fig. 50 d) with oval white scales in a pattern consisting of a median line, curved lateral lines and a line above the wing root. Pleuron with pattern of lines of broad white scales. Wing dark scaled; costa with a line of white scales on basal third or half. Sternites black with scattered white scales.

Adult Male

Palps slightly longer than proboscis, black, with white basal patches on segments 2-4; 5 almost entirely white scaled dorsally. *Terminalia* (fig. 50 a, b). Similar to that of *notoscriptus* but harpago with 2 strong setae on apical half. Lobes of IXth tergite with 1-3 setae.

Larva (fig. 50 f-j)

This can be distinguished from *notoscriptus* by the following traits: head seta 4, large, multibranched tuft. Lateral comb of 8-14 long spines in single row. Siphon slightly swollen at mid length; pecten of 17-23 spines.

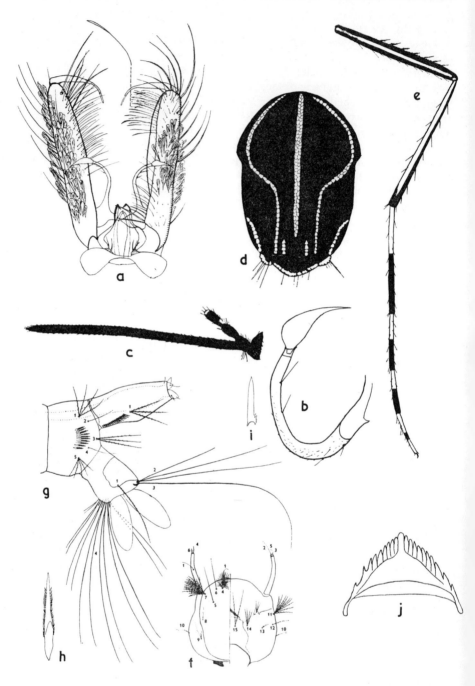

Fig. 50 *Aedes mallochi* Taylor. *a-e*, adult: *a*, male terminalia; *b*, harpago; *c*, female proboscis and palp; *d*, thorax; *e*, hind leg. *f-j*, larva: *f*, head; *g*, terminal segments; *h*, lateral comb tooth; *i*, pecten tooth; *j*, mentum. (*f-i*, after Marks.)

Anal segment: saddle without apical spines; seta 1 single; tufts of seta 4, 2 branched. Anal papillae equal, with rounded tips.

BIOLOGY

A tree-hole breeder. It has been recorded biting man just before and at dusk.

DISTRIBUTION

In Victoria the species is known from a single female collected at Underbool. It has also been recorded from Queensland and New South Wales.

Aedes (Finlaya) plagosus Marks

Aedes plagosus Marks, 1959, *Proc. Roy. Soc. Qd,* 70: 21.

A small black mosquito with broad white scales on the scutum forming longitudinal lines and a pair of patches. The proboscis is black, the tarsi banded. Though similar in some respects to *Ae. notoscriptus* and *Ae. mallochi,* it lacks the lyre-shaped pattern on the scutum.

ADULT FEMALE

Vertex black scaled with median triangular patch of narrow curved white scales; broad line of flat white scales along upper margin of eyes and turning back to the nape; upright forked scales black. Proboscis (fig. 51 a)

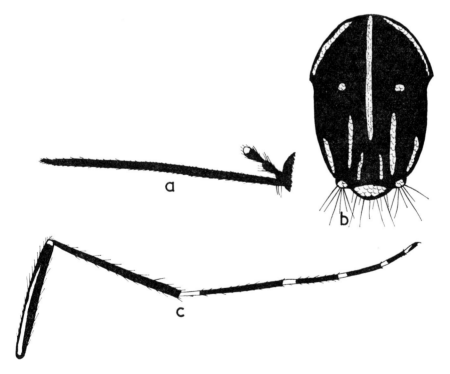

FIG. 51 *Aedes plagosus* Marks. *a,* female proboscis and palp; *b,* thorax; *c,* hind leg.

black scaled. Palps about quarter length of proboscis, black, with apical white patch. Scutum (fig. 51 b) black, clothed with narrow curved black scales with lines of silver scales; median line of narrow scales from anterior margin to prescutellar bare area; lines of broad scales along margin from anterior angle to scutal angle and three short lines lateral to prescutellar bare area; patch of broad scales lateral to dorsocentral bristles near mid length of scutum. Scutellum with broad band of silver scales on mid lobe and a patch on lateral lobes. Posterior pronotum with only a few dark narrow curved scales on upper margin. Postspiracular area with 2-3 setae. Broad flat silver scales form lines along paratergite and from subspiracular area across sternopleuron and middle of mesepimeron; patches below prealar knob, upper mesepimeron, lower sternopleuron and on propleuron. Wings dark scaled. Legs (fig. 51 c) black scaled: all femora with white kneespot and anterior line of white scales: mid tibia with anterior white line. Fore tarsi with two basal rings, mid tarsi with three, hind tarsi with four; last segment of hind tarsus white. Tergites black scaled with lateral basal silver patches. Sternites black scaled with lateral preapical silver patches.

MALE AND LARVA
Unknown.

BIOLOGY
A rare species, known from only two females. The holotype was collected in New South Wales and the second specimen off man at Maryborough, Victoria.

GROUP F (ALBOANNULATUS-GROUP)

This group has Neotropical, Nearctic and Palearctic representatives but the great majority of species occur in the Oriental and Australian Regions.

Characters of the Group

Wings dark scaled (except in two species). Scutal pattern does not consist of longitudinal narrow pale lines. Femora and tibiae without numerous spots or rings and without lines of white scales. Tarsi banded. Victorian species breed in ground and rock pools.

The only representatives of the group in southern Australia are members of the *alboannulatus* s. str. subgroup (Knight and Marks, 1952) and the taxonomy of the Victorian species has recently been clarified. Previously it was believed only two species occurred in Victoria: *Ae. queenslandis* (Strickl.) and *Ae. alboannulatus* (Macquart). However, a re-examination of the type of *Culex rubrithorax* Macquart by Klein and Marks (1960) has shown that it is an *Aedes* of the subgenus *Finlaya* and that it is conspecific with *Ae. queenslandis* (Strickland). Thus *queenslandis* becomes a synonym of *rubrithorax*. The recent discovery of three new species, *Ae. rupestris, Ae. tubbutiensis* and *Ae. subbasalis,* and a new record of *Ae. milsoni* have increased the number of species of this subgroup in Victoria to six.

KEY TO VICTORIAN SPECIES OF ALBOANNULATUS COMPLEX
Adults

1 Tibiae and proboscis mottled with pale scales 2
   Hind tibiae and proboscis not mottled 3
2(1) Scutum with patches of white scales. Femora with ochreous preapical ring.
   Venter white with median black patches . . . . *alboannulatus*
   No patches of white scales on scutum. No preapical ring on femora . . . . *tubbutiensis*
3(1) Tibiae with white sub-basal ring . . . . *subbasalis*
   Tibiae without sub-basal ring 4
4(3) Scutum with areas of white scales. Prescutellar area with broad white scales . . . . *milsoni*
   Scutum without areas of white scales. Prescutellar area with narrow scales 5
5(4) Venter more or less ochreous, usually with scattered black scales . . . . *rubrithorax*
   Venter black with white lateral patches. Apical border of sternites always with broad black band . . . . *rupestris*

Aedes (Finlaya) alboannulatus (Macquart)

Culex alboannulatus Macquart, 1849, *Dipt. exot.*, Suppl. 4: 10. *Culex alboannulatus* Theobald, 1901, *Mon. Cul.*, 1: 389. *Aedes alboannulatus* Edwards, 1924, *Bull. ent. Res.*, 14: 384. Dobrotworsky, 1959, 84: 132.

The proboscis is mottled; the scutum is clothed with dark-bronze scales and has areas of white scales; the femora have a preapical white band or patch; the tibiae are mottled; the tarsi are banded. The sternites are white scaled with black median patches.

ADULT FEMALE

Vertex clothed with narrow curved bronze scales except for small median area of white scales; upright scales black. Palps black scaled with white scales on base of all segments and on apex of last. Proboscis (fig. 52 d) black scaled with mottling, or pale scaled in middle. Scutal integument dark brown, almost black. Scutum clothed with narrow curved bronze scales with some small areas of white scales; a patch of broad white scales just in front of scutellum. Posterior pronotum with broad white scales below, broad black scales in middle and narrow bronze scales above; black area may have a few pale scales. Femora with preapical band or patch, which may join with knee spot. Tibiae mottled. Fore and mid tarsi with 2-3 white basal rings, hind tarsi with 4. Tergites black with incomplete white basal bands and lateral patches. Sternites white scaled with median, and lateral apical, black patches.

ADULT MALE

Palps about as long as proboscis without labella; with white spots at base of each segment. Proboscis less mottled than in female, sometimes with only a few white scales. Tergites II-VII with complete white basal bands; II may have only a patch of white scales; tergite VII may be mottled. Sternites usually black with lateral white spots; white scales may predominate and black scales be reduced to patches in middle of segment and apical corners. *Terminalia* (fig. 52 a-c). Coxite with scales and long and short seta, sternally and laterally; tergally clothed with numerous small setae. Basal lobe of coxite narrow, transverse, with a row of about 20 long setae along edge and some on upper surface. Style about half length of coxite, narrow, curved, with 2-3 preapical setae; terminal appendage long and straight. Harpago stout; appendage longer than harpago, not expanded in middle. Paraproct with single tooth. Lobes of IXth tergite prominent with 2-5 strong setae.

LARVA (fig. 52 f, g)

Head and siphon brown. Head about three-quarters as long as broad. Antenna about half length of head, spiculate; seta 1, 3-5 branched, arising at about mid length. Head setae: 4, 4-5 branched; 5, 3-5 branched; 6, 2-4 branched; 7, 6-7 branched; 8, 1-2 branched; 9, 2-3 branched. Head setae 4, 5 and 6 arranged to form apices of an almost right-angled triangle; seta 4 medial to 5 and 6 and between them. Mentum with 9-12 lateral teeth. Prothoracic setae: 1, 2 branched; 2, single, only slightly shorter than 1; 3, 4-5 branched, about three-quarters length of 2; 4, single, about as long as 3; 5 and 6, single; 7, 3 branched. *Abdomen*. VIIIth segment: comb large patch of fringed scales; seta 1, 3-5 branched; 3, 8-11 branched; 5, 4-6 branched; 2 and 4, single. Siphon index: 2·6-3·4, mean 2·9; seta 1, 6-7 branched; pecten of 17-20 spines, each with 3-4 teeth at base, central one the largest. Anal segment: saddle covering about half segment; seta 1, single or 2 branched; 2, 5-8 branched; 3, single; 4, of 15 tufts. Anal papillae about as long as saddle, or shorter.

BIOLOGY

Ae. alboannulatus is a sylvan species. It occurs almost throughout Victoria, but has not been recorded in the Mallee. It breeds in ground and rock pools; the water is usually clear but may be more or less cloudy. *Ae. alboannulatus* usually avoids heavily shaded pools in dense forest, showing a preference for diffuse sunlight. On the Bogong High Plains, at an altitude of 5,400 feet, it was found during the summer to be breeding in ground pools fully exposed to the sun. In the flat country north of the Dividing Range it can use such pools only during the cooler months of the year, being confined during the summer to shaded gullies, backwaters and roadside ditches.

The spring generation of *alboannulatus* oviposits on the edges of pools many of which will dry out during the summer; the eggs then remain dormant until autumn rains fill the pools again, when most of the eggs hatch within 12 hours. Some larvae pupate and may produce adults early in June, and these may oviposit later in the same month. In permanent water pools *alboannulatus* is able to breed all the year round.

Adults are always abundant during the spring, and commonly during the autumn in favourable places, but a prolonged dry summer may result

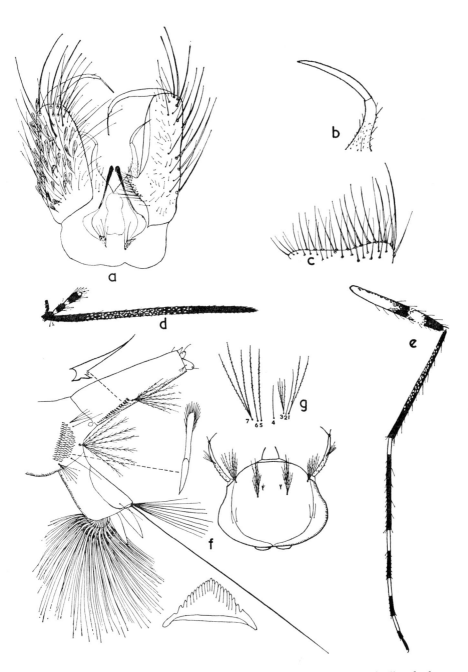

Fig. 52 *Aedes alboannulatus* (Macquart). *a-e*, adult: *a*, male terminalia; *b*, harpago; *c*, basal lobe of coxite; *d*, female proboscis and palp; *e*, hind leg. *f-g*, larva: *f*, head, mentum and terminal segments; *g*, prothoracic setae.

in the drying out of pools which in normal years permit continuous breeding.

The adults are vicious day-biting mosquitoes; they bite even during the winter at temperatures as low as 52°F.

DISTRIBUTION

Ae. alboannulatus is widely distributed in Victoria but is apparently absent from the north-western part of the state. It occurs in all the eastern states of Australia, and in Tasmania, South Australia and southern Western Australia.

Aedes (Finlaya) rubrithorax (Macquart)

Culex rubrithorax Macquart, 1850, *Dipt. exot.,* Suppl. 4: 9. *Culicelsa queenslandis* Strickland, 1911. *Entom.,* 44: 179. *Culicelsa similis* Strickland, 1911, ibid., 44: 132. *Culicada demansis* Strickland, 1911, ibid., 44: 202. *Culicada cumpstoni* Taylor, 1914, *Trans. ent. Soc. Lond.,* 1913: 692. *Culicada hybrida* Taylor, 1916, *Proc. Linn. Soc. N.S.W.,* 41: 568. *Aedes rubrithorax* Klein and Marks, 1960, *Proc. Linn. Soc. N.S.W.,* 85: 108. Dobrotworsky, 1959, 84: 134.

A member of the *alboannulatus* subgroup. The proboscis is black scaled; the scutum is without areas of white scales; there is no preapical white band on the femora and the tibiae are usually dark. The sternites have some scales with an ochreous tint and some scattered black ones.

ADULT FEMALE

Vertex usually with large patch of golden yellow scales. Proboscis black scaled. Thoracic integument usually light brown. Scutum clothed with narrow curved golden-yellow scales, mixed laterally, particularly in the area near the scutal angle, with dark-bronze scales. Prescutellar area without broad scales. Posterior pronotum with patch of broad white scales below, broad black scales in middle and golden narrow curved scales above; area of narrow golden scales reduced in some specimens to a line, but in others it extends downwards, reducing the black area. Pleuron with patches of broad creamy scales and bristles. Fore and mid femora black with mottling of yellowish scales, except for ventral side of basal two-thirds, which is pale scaled. Hind femora (fig. 53 c) pale on basal half, black with yellowish mottling on apical half; apical part of femur sometimes yellow. Tibiae black. Fore and mid tarsi with white basal bands on first two or three segments; hind tarsi with bands on first four segments and sometimes with narrow band on fifth. Tergites black scaled with white lateral spots and incomplete yellowish basal band on tergites II-VI; in some specimens the bands are reduced to a few pale scales. Tergites VII-VIII black, or mottled with yellowish scales; in some specimens yellow scales predominate on tergites VI-VIII. Sternites clothed with ochreous scales, with apical lateral black spots and usually mottling of black scales; black scales may predominate.

Adult Male

Palps black scaled, as long as proboscis without labella; segments 2-5 each with patch of white scales. Pronotum as in the female or predominantly black scaled. Sternites sometimes with mixture of black, yellow and white scales, or white with black median and apical lateral patches. *Terminalia* (fig. 53 a, b). Coxite clothed basally with white scales and apically with black scales and golden bristles. Style about half length of coxite, narrowing to both ends, with 1-2 preapical setae; terminal appendage slightly curved and about half length of style. Basal lobe of coxite narrow, transverse, with row of about 10 setae. Harpago stout, with fine setae at base; appendage about as long as harpago, widened at middle and narrowing at end. Paraproct with single tooth and 5 fine setae. Lobes of IXth tergite prominent with 2-7 stout setae.

Larva (fig. 53 d, e)

Head and siphon light brown. Head about three-quarters as long as broad. Antenna about half length of head, spiculate; seta 1, 4-6 branched, about half length of antenna and arising about mid length. Head setae; 4, tiny, 4-6 branched; 5, 4-9 branched; 6, 3-5 branched; 7, 8-10 branched; 8, single; 9, 2-3 branched. Setae 4, 5 and 6 arranged to form apices of triangle; setae 4 arising medial to setae 5, slightly in front of, or slightly behind, a line drawn through them. Mentum with 12-13 lateral teeth. Prothoracic setae: 1, 2 branched; 2, single; 3, 5-10 branched, about three-fifths as long as 2; 4, small, 2 branched or single; 5 and 6, single; 7, 3 branched. *Abdomen*, VIIIth segment: lateral comb large patch of about 200 fringed scales; seta 1, 4-6 branched; 3, 8-13 branched; 5, 4-9 branched; 2 and 4, single. Syphon index $3 \cdot 1$-$3 \cdot 7$, seta 1, 8-9 branched; pecten of 19-36 spines, each with 4-5 teeth at base. Anal segment: saddle covering about half segment; seta 1, 1-2 branched; 2, 6-11 branched; 3, single, long; 4, of 13-16 tufts. Anal papillae about as long as saddle.

Yellow Form

This form differs from other variants in its general yellow colour, but although these are quite distinct from the type form of *rubrithorax* there are intermediates.

The yellow form is characterized by its lighter integument and the following features. Proboscis with a few pale scales on basal half; torus yellow, darker on inner side. Broad scales on lower part of posterior pronotum yellowish. Fore and mid femora intensively mottled with yellowish scales, hind femora yellowish with mottling of black scales. Fore tibiae pale below, black above; mid tibia mottled with yellowish scales; hind tibia black above and apically, elsewhere yellow. Tarsi of fore and mid legs with 3 banded segments, hind with 4; band on segment 4 about quarter length of segment. Bands on tergites in females, reduced to yellow patches, or absent; last 2-3 tergites clothed with ochreous scales. Sternites clothed with ochreous scales and a few scattered black ones; apical lateral spots black.

Biology

Ae. rubrithorax is confined to woodlands. It breeds in swamps, ground

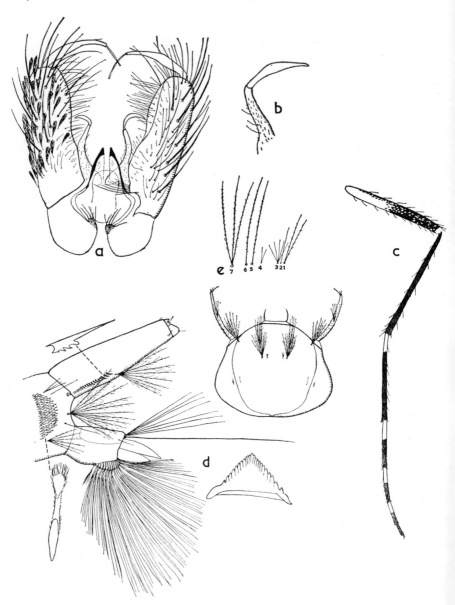

Fig. 53 *Aedes rubrithorax* (Macquart). *a-c*, adult: *a*, male terminalia; *b*, harpago; *c*, hind leg. *d-e*, larva: *d*, head, mentum and terminal segments; *e*, prothoracic setae.

and rock pools, dams, cavities in logs, and in artificial containers (tanks, tins, etc.), usually in shaded situations since it requires water of relatively low temperature. It may make use of exposed shallow pools during the cooler months of the year, but in the summer can do so only at high altitudes, where deep pools exposed to the sun are always suitable.

In Victoria the temperature of the water in which *Ae. rubrithorax* breeds has been found not to exceed 65°F, even during the summer. During the coldest months development of larvae is slow, but a few pupae can be found almost throughout the winter at lower altitudes. In the laboratory, emergence of adults was observed at 51-52°F. *Ae. rubrithorax* oviposits on moist soil or rock surfaces just above water level. The eggs cannot withstand severe desiccation, but in natural conditions, in forests, they may remain viable throughout the summer and hatch after autumn rains fill the pools. It is a common man-biting bush mosquito.

DISTRIBUTION

In Victoria *Ae. rubrithorax* is usually restricted to woodlands and has its northern distribution to the isohyet of 20 inches. In other states its distribution is similar to that of *Ae. alboannulatus* but it is absent from Western Australia.

Aedes (Finlaya) rupestris Dobrotworsky

Aedes rupestris Dobrotworsky, 1959, *Proc. Linn. Soc. N.S.W.*, 84: 136.

A member of the *alboannulatus* subgroup. Rather similar to *Ae. rubrithorax*. The sternites are black scaled with basal lateral white patches or white scaled with a median patch and an apical border of black scales.

ADULT FEMALE

Vertex with narrow curved scales, yellow goldish in centre; upright forked scales yellow-golden, becoming dark towards sides and neck. Palps black scaled with patch of white scales on segment 2 and at base and apex of segment 3. Scutal integument dark brown. Scutum clothed with narrow dark-golden scales, becoming pale around bare area, in front of wing roots, along border with posterior pronotum and anteriorly; admixture of black scales laterally. Posterior pronotum with patches of broad white scales below, broad black scales in middle and narrow curved goldish scales above. Pleuron with patches of broad white scales and pale bristles. Leg (fig. 54 c) fore and mid femora mottled, pale posteriorly on basal half; tarsal segments 1 and 2 with white bands, 3 with a few white scales at base. Hind femur pale ventrally on basal two-thirds; apical third black scaled with few white scales posteriorly. Four basal hind tarsal segments banded; band on segment 4 half length of segment. All legs with creamy knee spot and black tibia. Tergites black scaled with white lateral patches on all segments; II-V with incomplete narrow white basal bands; VI and VII each with patch of pale scales. Sternites black scaled with white lateral patches; apical border always with complete broad black band.

ADULT MALE

Palps black scaled, as long as proboscis without labella; last two segments and apex of shaft with long dark hairs; segments 2-5 each with patch of white scales. Tergite I with few dark and pale scales in middle, II with an incomplete basal band, III-VI with basal bands joining white lateral spots, VII with a few white scales at base, VIII with lateral white

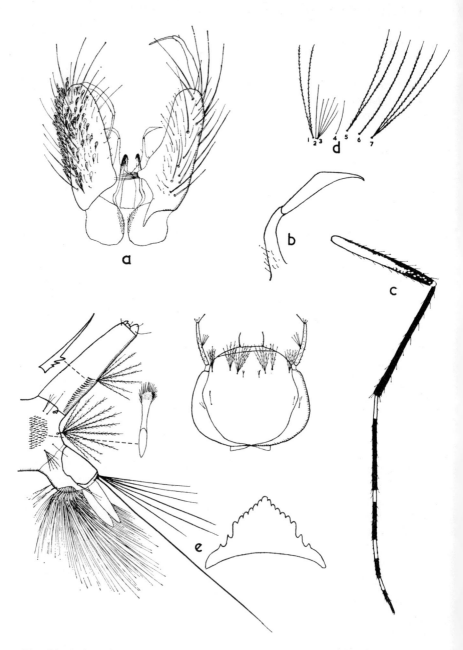

Fig. 54 *Aedes rupestris* Dobrotworsky. *a-c*, adult: *a*, male terminalia; *b*, harpago; *c*, hind leg. *d-e*, larva: *d*, prothoracic setae; *e*, head, mentum and terminal segments.

spots only. Sternites black with elongate white lateral spots not reaching apical edge. *Terminalia* (fig. 54 a, b). Coxite black scaled, with patch of white scales basally; laterally and sternally it bears long, and some short, setae. Style half length of coxite, narrowing sharply at mid length; 1-2 preapical setae; terminal appendage straight, about half length of style. Basal lobe of coxite narrow, transverse, with row of about 10 long setae along edge and several small ones on upper side. Harpago stout, with fine setae at base; appendage about as long as harpago, widened at middle and narrowing to end. Paraproct with single tooth and 2 very fine setae near tip. IXth tergite without prominent lobes; a few fine short setae on inner side.

LARVA (fig. 54 d, e)

Head and siphon dark brown, body blackish. Head about four-fifths as long as broad. Antenna about two-fifths length of head, spiculate; seta 1 with 2-4 branches, less than half length of antenna and arising at about two-fifths length from base. Head setae: 4, tiny, 4-7 branched; 5, 5-7 branched; 6, 3-5 branched; 7, 5-10 branched; 8, single; 9, 2-3 branched. Setae 4, 5 and 6 with bases almost in a straight line. Mentum with 8-10 lateral teeth on each side. Prothoracic setae: 1, 2 branched, long; 2, single, about three-fifths length of seta 1; 3, stellate, 5-11 branched and about half length of seta 1; 4, short, 1-2 branched; 5, long, 2-3 branched; 6, single, as long as 5; 7, 3 branched. *Abdomen.* VIIIth segment: lateral comb of about 150 fringed scales; seta 1, 3-4 branched; 3, 7-13 branched; 5, 5-7 branched; 2 and 4, single. Siphon stout, index 2·6-3·5, mean 2·9; seta 1, 6-10 branched; pecten of 16-28 spines each with 2-3 teeth at base, the upper the largest. Anal segment: saddle covering more than half the segment; seta 1, usually 2 branched, sometimes single; 2, 6-10 branched; 3, single, long; 4 of 14 tufts, one or two precratal. Anal papillae three-fifths to four-fifths length of saddle.

BIOLOGY

The breeding places of *Ae. rupestris* in Victoria are exposed rock pools in eucalypt forests. The pools are usually small and shallow with a thick layer of black mud and decayed leaves on the bottom. The water is brown, with a strong smell of decomposing eucalypt leaves. The temperature of the water in such pools often rises during the summer to 86-93°F. In south Queensland the larvae of this species have been collected, in large numbers, in exposed rock pools in which water temperature in the early afternoon exceeded 99°F.

In some rock pools, particularly shallow ones with sloping edges, *Ae. rupestris* alone was found during the summer, but in these same pools there were always some *rubrithorax* larvae from late autumn to early spring. In large and deep pools with cleaner and cooler water, or with some vegetation, *rubrithorax* was always more numerous than *rupestris* even during the summer.

The eggs, which remain dormant on the edges of dried pools, hatch after being submerged by later rains. In the laboratory, hatching of eggs was observed at temperatures in the range of 49-58°F. At such low temperatures the larvae developed only slowly, but eventually produced adults. Successful emergence of adults was observed in the laboratory at temperatures as low as 50°F.

It is a day-biting mosquito, and is very common near its breeding sites; it ceases biting early in the winter when the temperature falls to 62-63°F.

DISTRIBUTION

Victoria: Franklin River, Little River, W Tree Creek (north from Buchan), Tubbut, Nowa Nowa, all in Gippsland, Warburton, Lorne, Meredith, Grampians. It has also been recorded from Queensland.

Aedes (Finlaya) tubbutiensis Dobrotworsky

Aedes tubbutiensis Dobrotworsky, 1962, *Proc. Linn. Soc. N.S.W.*, 84: 139.

A member of the *alboannulatus* subgroup. Very similar to *Ae. rubrithorax*. Distinguished by having the proboscis and the tibiae mottled with white scales. The sternites are white without an ochreous tint.

ADULT FEMALE

Vertex with pale narrow, curved scales; upright forked scales in centre pale. Palps black scaled with patch of white scales at base and at apex of segment 3; segments 2 and 3 with a few pale scales above. Proboscis black with mottling of white scales on basal two-thirds. Scutal integument almost black. Scutum clothed with narrow bronze and black scales, becoming pale around bare area, near wing roots and on margins of scutum; two lateral pale patches near mid length; scales in front of scutellum narrow. Posterior pronotum with patch of broad white scales below, broad black scales in middle and narrow curved pale scales above; a few pale broad scales scattered in upper part of black area. Pleura with usual patches of broad white scales and pale bristles. Femora black scaled with white mottling; basal half pale scaled posteriorly. Knee spots yellowish. Tibiae black, mottled with white scales. Tarsal segments 1-2 of fore legs, 1-3 of mid legs and 1-4 of hind legs with basal white bands; band on segment 4, half length of segment. Tergites black scaled. Tergite I with few pale and black scales in middle, II-IV with incomplete basal white bands and lateral spots, V-VII with bands joining lateral spots. Tergites VI-VIII with scattered pale scales, increasing towards segment VIII. Sternites white scaled with black mottling in middle and small apical, lateral spots.

ADULT MALE

Proboscis black with occasional pale scales. Palps about as long as proboscis with labella; segments 2-5 with basal patches of white scales. Tarsal segments 1-2 of fore legs, 1-3 of mid legs and 1-4 of hind legs with white basal bands. Tergite II with narrow basal band, III-VI with wide basal bands joining lateral spots. Sternites black scaled, with elongate, white lateral spots and a few white scales in middle of apical border. *Terminalia* (fig. 55 a, b). Very similar to that of *Ae. rubrithorax* and *Ae. rupestris*, but the style is shorter—about a third of the length of the coxite. The first 4-5 setae of the basal lobe are about twice as long as those in *rubrithorax*.

SUBGENUS FINLAYA THEOBALD

LARVA (fig. 55 d, e)

Head and siphon dark brown; body blackish. Head about three-quarters as long as broad. Antenna about half length of head, spiculate; seta 1, 4-5 branched, arising about mid length. Head setae 4, tiny, 5-7 branched; 5, 5-7 branched; 6, 3-5 branched; 7, 7-10 branched; 8, single, rarely forked at apex; 9, 2-3 branched, Mentum with 10-12 lateral teeth on each side. Prothoracic setae: 1 and 2, single; 3, 4-6 branched, as long as 2; 4, short, 2 branched; 5 and 6, single; 7, 3 branched. *Abdomen.* VIIIth

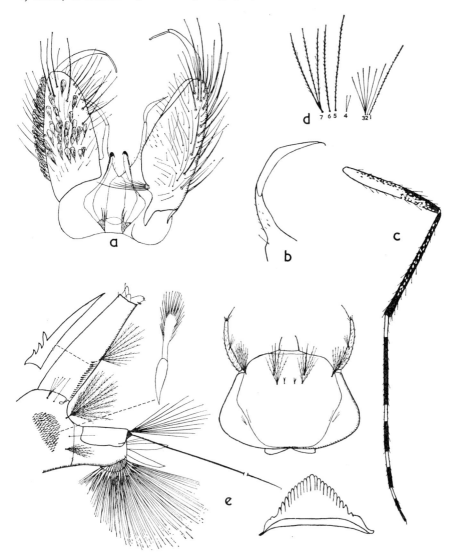

FIG. 55 *Aedes tubbutiensis* Dobrotworsky. *a-c*, adult: *a*, male terminalia; *b*, harpago; *c*, hind leg. *d-e*, larva: *d*, prothoracic setae; *e*, head, mentum and terminal segments.

segment: lateral comb a large patch of fringed scales; seta 1, 4-5 branched; 3, 11-12 branched; 5, 6-7 branched; 2 and 4, single. Siphon index 3·1-3·7, mean 3·4; seta 1, 7-11 branched, pecten of 20-30 spines, mean 25. Anal segment: saddle covering about half segment; seta 1, single, may be 2 branched; 2, 6-12 branched; 3, single; 4 of 15-16 tufts. Anal papillae about as long as saddle.

BIOLOGY

Ae. tubbutiensis breeds mainly in back waters in creek beds, and at times in rock pools more or less exposed to the sun. It avoids completely shaded pools. Adults have not been collected in nature and there is no information on their biting habits. It can be expected that it will bite man as readily as do other Victorian members of this complex.

DISTRIBUTION

All known breeding places are situated in the north-eastern corner of Gippsland, in wooded country at an elevation of about 1,800 feet. This area has an average rainfall of 25-30 inches.

Aedes (Finlaya) subbasalis Dobrotworsky

Aedes subbasalis Dobrotworsky, 1962, *Proc. Linn Soc. N.S.W.*, 87: 298.

A member of the *alboannulatus* subgroup. Similar to *Ae. rupestris*. Easily distinguished by having a ring or a patch of white scales near the base of the tibiae.

ADULT FEMALE

Narrow curved white scales on vertex, forming a broad triangular patch sometimes reduced to a narrow median patch. Upright forked scales black. Proboscis black. Scutal integument dark almost black. Scutum clothed mostly with bronze scales with golden reflections, or white scales may be present on front and lateral margins, around bare area, and may form two short dorsocentral lines. Scutellum with narrow white scales. Posterior pronotum with broad white scales below, narrow curved black scales in middle and narrow curved white scales above. Pleura with patches of broad white scales and pale bristles. Fore and mid femora black mottled with white scales; hind femora white with black mottling on basal half; black with white mottling on distal half. Tibiae (fig. 56 d) black scaled with incomplete subbasal ring. Fore tarsi with two broad white basal bands, mid tarsi with three, hind tarsi with four. Knee spots white. Tergites with constricted basal white bands. Sternites white scaled with black apical bands joined to elongate patches of black scales. Upper fork cell about twice the length of its stem.

ADULT MALE

Palps slightly shorter than proboscis without labella, black scaled with white patches at base of segments 2-5. *Terminalia* (fig. 56 a-c). Coxite black scaled with some white scales basally, long and short setae laterally. Basal lobe of coxite narrow, transverse, with row of about 10 long setae along edge. Style about half length of coxite, narrow, curved; terminal

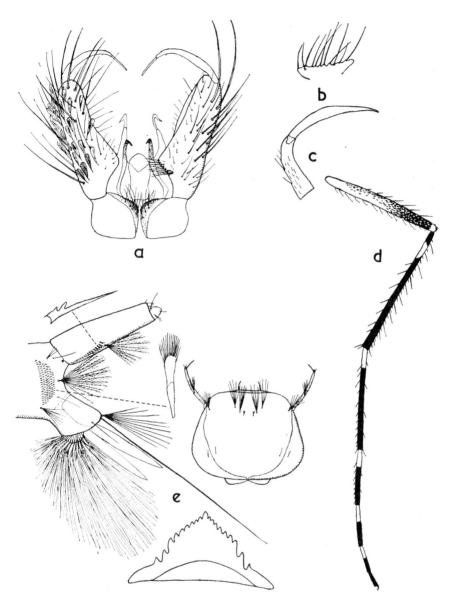

FIG. 56 *Aedes subbasalis* Dobrotworsky. *a-d*, adult: *a*, male terminalia; *b*, basal lobe; *c*, harpago; *d*, hind leg. *e*, larva: head, mentum and terminal segments. (*a, c, e*, after Dobrotworsky.)

appendage straight and long. Harpago stout; appendage longer than harpago. Paraproct with single tooth. Lobes of IXth tergite large, with 3-8 setae.

LARVA (fig. 56 e)

Head only slightly wider than long. Antenna about half length of head.

Head setae: 4, small, 3-5 branched; 5 and 6, 5-6 branched; 7, 6-9 branched; 8, single; 9, 2-3 branched. Setae 4, 5 and 6 with bases almost in a straight line. Mentum with 9-11 lateral teeth on each side. Prothoracic setae: 1 and 5, 2 branched; 2 and 6, single; 3, 5 branched; 4 and 7, 3 branched. *Abdomen*. VIIIth segment: lateral comb of more than one hundred scales; seta 1, 2-5 branched; 2 and 4, single; 3, 9-12 branched; 5, 4-6 branched. Siphon index from 3·0 to 3·6, mean 3·3; seta 1, 7-11 branched; pecten of 17-21 spines. Anal segment: saddle covering about one-third of the segment. Seta 1, single, rarely 2 branched; 2, 19-12 branched; 3, single; 4 (ventral brush) of 14-16 tufts. Anal papillae long, slightly less than twice length of saddle.

BIOLOGY

The breeding places of *Ae. subbasalis* are rock pools in river and creek beds which are usually exposed to sun but may be partly shaded. The water may be pale brown or clear, with decayed leaves and debris on the bottom. It is a day-biting mosquito often attacking man in bright sunlight.

DISTRIBUTION

In Victoria *Ae. subbasalis* occurs in East Gippsland (Weeragua, Tubbut, Little River). It has also been recorded from New South Wales and Queensland.

Aedes (Finlaya) milsoni Taylor

Culicada milsoni Taylor, 1915, *Proc. Linn. Soc. N.S.W.*, 40: 179. *Hulecoeteomyia milsoni* Taylor, ibid., 41: 566. *Aedes milsoni* Edwards, 1924, *Bull. ent. Res.*, 14: 385. Dobrotworsky, 1962, 87: 301.

A member of the *alboannulatus* subgroup. This species has patches of white scales on the scutum and broad white scales on the prescutellar area. The sternites are white with apical black bands.

ADULT FEMALE

Vertex with narrow curved white scales; upright scales mostly black with some pale ones medially. Proboscis black scaled. Scutal integument dark brown. Scutum (fig. 57 d) clothed with narrow curved dark-bronze or light-brown scales; there are white scaled areas on margins; prescutellar area with patches of broad white scales. Posterior pronotum with broad black scales in middle, narrow curved pale scales above and broad white scales below. Scutellum with narrow white scales. Femora (fig. 57 e) mottled anteriorly, knee spots white. Tibiae black. Fore and mid tarsi with 2-3 basal white bands, hind tarsi with four. Tergites black scaled with incomplete white basal bands and lateral patches. Sternites white scaled, with apical black band and black median patch.

ADULT MALE

Palps black scaled, almost as long as proboscis with labella; segments 2-5 with patch of white scales. Tergites with complete white basal bands. Sternites black with white basal lateral patches. *Terminalia* (fig. 57 a-c). Coxite with black scales and setae. Basal lobe of coxite narrow, transverse, with a row of about 14 long setae. Style about one-third length of coxite,

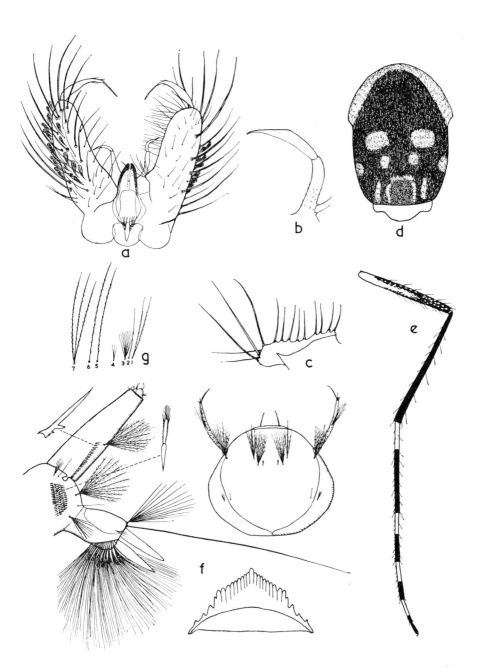

Fig. 57 *Aedes milsoni* Taylor. *a-e*, adult: *a*, male terminalia; *b*, harpago; *c*, basal lobe of coxite; *d*, thorax; *e*, hind leg. *f-g*, larva: *f*, head, mentum and terminal segments; *g*, prothoracic setae. (*f*, after Dobrotworsky.)

narrowing to both ends; terminal appendage long and almost straight. Harpago stout; appendage about as long as harpago. Paraproct with single tooth. Lobes of IXth tergite prominent, with 3-4 setae.

LARVA (fig. 57 f, g)

Head only slightly wider than long. Antenna short, about half length of head; seta 1, 5-7 branched: Head setae: 4, small, 5 branched; 5, 7-10 branched; 6, 3-5 branched, usually with one branch much thicker than others; 7, 12-15 branched; 8, single; 9, 4-5 branched. Mentum with 11-13 lateral teeth on each side. Prothoracic seta 1, 2 branched; 2, 5 and 6, single; 3, 7-12 branched; 4 and 7, 3 branched. *Abdomen*. VIIIth segment: lateral comb of more than one hundred scales; seta 1, 5-6 branched; 2 and 4, single; 3, 11-13 branched; 5, 6-7 branched. Siphon index 3·0-3·3; pecten of 25-26 spines; seta 1, 9-10 branched. Anal segment: seta 1, 3 branched; 2, 12-13 branched; 3, single; 4 (ventral brush) of 14 tufts. Anal papillae slightly longer than length of saddle.

BIOLOGY AND DISTRIBUTION

In Queensland *Ae. milsoni* breeds in partly shaded ground and rock pools, with leaves on the bottom and with clear or discoloured water.

It is a northern species recorded in Queensland and in New South Wales. In Victoria it has been collected only at Maryborough in September and November.

GROUP UNDETERMINED

Aedes (Finlaya) subauridorsum Marks

Aedes subauridorsum Marks, 1948, *Pap. Dep. Biol. Univ. Qd,* 2 (8): 28.

The scutum is clothed with bright golden scales. The basal half of the hind femur is white; white bands are present only on the first, or first and second, tarsal segments.

ADULT FEMALE

Vertex with narrow curved and upright golden scales. Palps and proboscis black scaled. Integument dark brown. Scutum clothed with narrow curved bright golden scales. Posterior pronotum with a few lanceolate golden scales. Patches of broad silvery scales on propleuron, anterior pronotum, paratergite, sternopleuron and mesepimeron. Wings and legs with purplish-black scales. Fore and mid femora white scaled below on basal two-thirds hind femur (fig. 58 B) with basal two-thirds entirely white scaled; knee spot large. First tarsal segment of fore and mid tarsi with inconspicuous basal band, of hind tarsi with conspicuous band. Abdomen purplish-black scaled; tergites without bands but with large lateral patches of silvery-white scales. Sternites with narrow basal bands joining large lateral patches.

ADULT MALE

Palps almost as long as proboscis. Posterior pronotum with whitish broad scales basally and elongate scales above. *Terminalia* (fig. 58 A). Coxite black scaled, about 3½ times as long as broad; basal lobe slightly developed with patch of about 30 mesially directed setae and sternally, on apical third, a row of 15-20 long fine curved setae. Style short, slightly curved, tapering; terminal appendage long. Harpago stout with long preapical seta; appendage broad at base, tapering to curved pointed tip. Paraproct with 1-2 teeth. Lobes of IXth tergite with 4-11 setae.

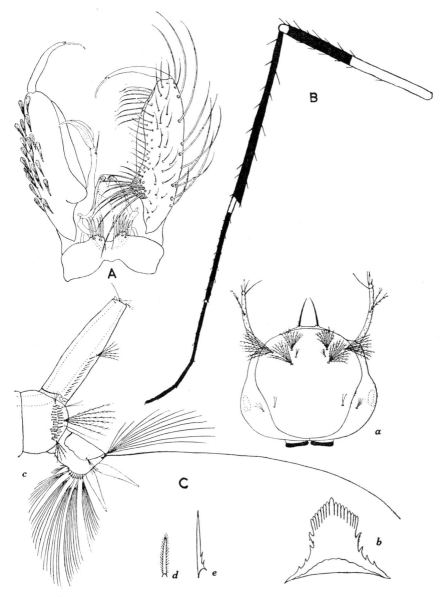

FIG. 58 *Aedes subauridorsum* Marks. A-B, adult: A, male terminalia; B, hind leg. C, *a-e*, larva: *a*, head; *b*, mentum; *c*, terminal segments; *d*, lateral comb tooth; *e*, pecten tooth. (A, C, after Marks.)

LARVA (fig. 58 Ca-e)
Head and saddle light brown, siphon darker, particularly towards apex. Antenna short; seta 1, single or 2 branched. Head setae: seta 1, long; 4, small tuft; 5 and 6, 6-10 branched; 7, 7-10 branched; 8, 2 branched; 9, 3-6 branched. Mentum long and narrow, with longer central tooth and 10-13 lateral teeth on each side. *Abdomen.* VIIIth segment: lateral comb of 19-31 narrow fringed scales in 2-3 rows; seta 1, 3-4 branched; 2, 2-3 branched; 3, 6-8 branched; 4, single; 5, 4-6 branched. Siphon index about 3·5; pecten of 10-16 spines; seta 1, 6-7 branched. Anal segment: saddle covering dorsal half of segment; seta 1, 4-5 branched; 2, 10-12 branched; 3, single; 4, of 12 tufts. Anal papillae equal, pointed, about as long as saddle.

BIOLOGY
Ae. subauridorsum breeds in tree holes. It is a man-biting species and attacks during the day.

DISTRIBUTION
In Victoria only 2 specimens have been collected: the male at Woods Point and the female at Maroondah. It has also been recorded from Queensland and New South Wales.

Subgenus MACLEAYA Theobald

Macleaya Theobald, 1903, *Entom.*, 36: 154.

This subgenus was erected for a single species *Ae tremulus* and, though apparently related to *Finlaya,* clearly differs from it. The second species *Ae. littlechildi* has been described from New Guinea by Taylor.

Characters of the Subgenus

Vertex with broad and flat scales. Palps of male about as long as proboscis, last two segments rather short and turned slightly downwards. No lower mesepimeral bristles. All claws of female simple. No tuft of long scales at extreme base of wing. Abdomen of female rather stout, somewhat depressed; VIIIth segment small and almost hidden; cerci short and broad. Male terminalia: coxite simple, basal lobe imperfectly separated, hairy; harpago with long but not expanded appendage.

Aedes (Macleaya) tremulus (Theobald)

Macleaya tremula Theobald, 1903, *Entom.*, 36: 154. *Danielsia minuta* Taylor, 1912, *Bull. Northern Terr.*, la: 30. *Danielsia alboannulata* Taylor, 1912, ibid.: 31. *Aedimorphus australis* Taylor, 1914, *Proc. Linn. Soc. N.S.W.*, 39: 457. *Aedimorphus australis* var. *darwini* Taylor, 1914, ibid.: 458. *Mimeteomyia doddi* Taylor, 1919, ibid., 43: 831. *Aedes tremula* Edwards, 1924, *Bull ent. Res.*, 14: 385. *Aedes tremulus* Stone, Knight and Starke, 1959, *Syn. Cat. Mosquitoes World*, 6: 167.

A small rather stoutly-built mosquito. The scutum is clothed with black scales and a few scattered white ones. The tarsi have white basal bands on the first three segments; on the hind tarsus the fourth segment is all black, the fifth all white.

ADULT FEMALE

Vertex with broad flat black scales; some narrow curved white scales medially. Palps black with white patch at tip. Proboscis black scaled. Scutum clothed with narrow black scales, and a few scattered white scales becoming more numerous round bare area. Posterior pronotum dark scaled with some white scales below. Pleuron with patches of broad white scales. Wing black scaled; no scale-tuft at base of wing. Legs (fig. 59 c) black; knee spots small. Fore and mid tarsi with white basal bands on segments 1 and 2; hind tarsi with bands on segments 1, 2 and 3; segment 5, white. All claws simple. Abdomen black scaled with small basal median patches, and larger lateral patches, of white scales. Venter pale scaled basally, black apically.

ADULT MALE

Palps (fig. 59 b) about as long as proboscis with labella, dark scaled; last two segments rather short, the penultimate white at base, the terminal all white. *Terminalia* (fig. 59 a). Coxite simple, without basal or apical lobes; coxite with patch of scales laterally, short setae on basal two-thirds and long setae on apical third; base of coxite on inner side, with longitudinal row of long setae beyond which are numerous flattened, mesially directed setae. Style about one-third length of coxite, slender; appendage about half length of style, slightly curved. Harpago with long rod-like appendage. Paraproct with single tooth. Lobes of IXth tergite with 3-4 strong, long setae.

LARVA (fig. 59 d)

Head almost round. Antenna almost as long as head; seta 1, single. Head setae: 4, a tuft; 5, 2 branched; 6, curved, 3 branched; 7, 2-5 branched; 8, single; 9, 2-3 branched. *Abdomen.* VIIIth segment: lateral comb row of 3-5 strong spines with fused sclerotized bases; setae 1, 2 and 4, single; 3, 5 branched; 5, 2 branched; siphon short, tapering, index about 1·5; pecten of about 12 spines; seta 1, single. Anal segment: saddle reduced; setae 1 and 2, single; 3, 4-5 branched; 4, 2 branched; anal papillae large, broad with blunt tips, about 4 times as long as saddle.

BIOLOGY

Larvae have been found in tree holes, water tanks, septic tanks and wells. It is a man-biting mosquito.

DISTRIBUTION

In Victoria, *Ae. tremulus* has been recorded only from the Mildura area and from Maryborough. It is widely distributed in other parts of Australia and also occurs in New Guinea.

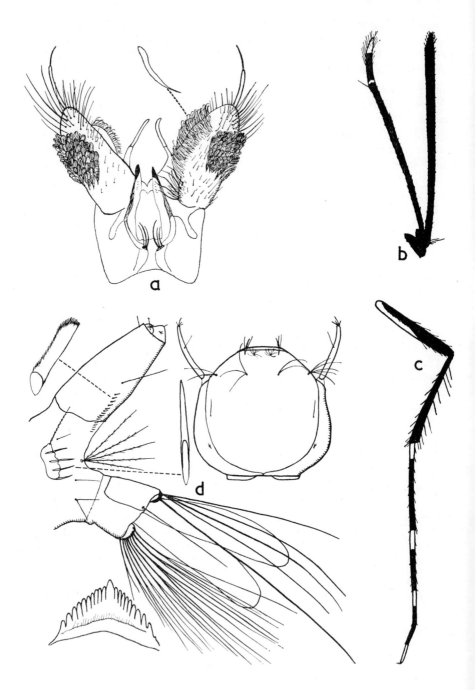

Fig. 59 *Aedes tremulus* (Theobald). *a-c*, adult: *a*, male terminalia; *b*, male proboscis and palp; *c*, hind leg. *d*, larva: head, mentum and terminal segments.

Subgenus CHAETOCRUIOMYIA Theobald

Chaetocruiomyia Theobald, 1910, *Mon. Cul.*, 5: 196.

This small subgenus is endemic to Australia. From eight described species only two occur in Victoria. Apparently it is closely related to the subgenus *Macleaya*, but its relationships to this and other subgenera will remain uncertain until the males and larvae of more species have been described.

Characters of the Subgenus

Small but rather well-ornamented species. Scales of vertex, pronotal lobes and scutellum broad and flat. Proboscis rather short. Legs rather short and thick; dorsal bristles on tibiae usually long. Wings with a tuft of very long scales on the small sclerotized piece at the extreme base (fig. 60 b). Abdomen short and rather stout; cerci short and hidden.

Aedes (Chaetocruiomyia) wattensis Taylor

Aedes wattensis Taylor, 1929, *Bull. ent. Res.*, 20: 275.
Marks, 1964, 89: 144-5.

A very small (wing length 1·4-1·9 mm) stoutly-built mosquito with a short proboscis. It is easily recognized by the fact that the anterior half of the scutum is pale, the posterior half dark; the hind margin of the pale area has a median indentation. The vertex is white scaled medially and black laterally. The tarsi are banded.

Adult Female

Vertex with broad white scales; upright forked scales dark; lateral scales black. Proboscis and palps black scaled. Scutal integument dark brown. Scutum (fig. 60 a) clothed with pale scales on anterior half; posterior margin of pale area distinctly indented between dorsocentral bristles. Posterior half of scutum clothed with dark-bronze scales. Posterior pronotum with dark scales and occasionally a few pale scales. Wing dark scaled, some white scales at base of C. Fore and mid femora black with white patch posteriorly towards middle; hind femora (fig. 60 c) black with white knee patch. All tibiae black. Fore and mid tarsi with two basal white bands; hind tarsi with three basal white bands, fifth segment all white. Tergites black scaled with median basal and lateral white patches, VIIIth sometimes with basal white band. Sternites white scaled; V-VI sometimes dark apically.

Adult Male

Palps longer than proboscis without labella by segment V. *Terminalia* (fig. 60 d). Coxite bears a number of long setae with dense scales laterally, extending onto mid third tergally. Basal lobe extending two-thirds length of coxite, densely clothed with moderately long setae; on distal margin of lobe there are 1-2 long setae followed by a row of 5-6 broad striated scale-like dark setae, and 5-6 smaller striated setae; along distal half to two-thirds mesially there are setae with very broad, curved, flattened tips, and lateral to these are 1-2 rows of scales with narrower flattened tips.

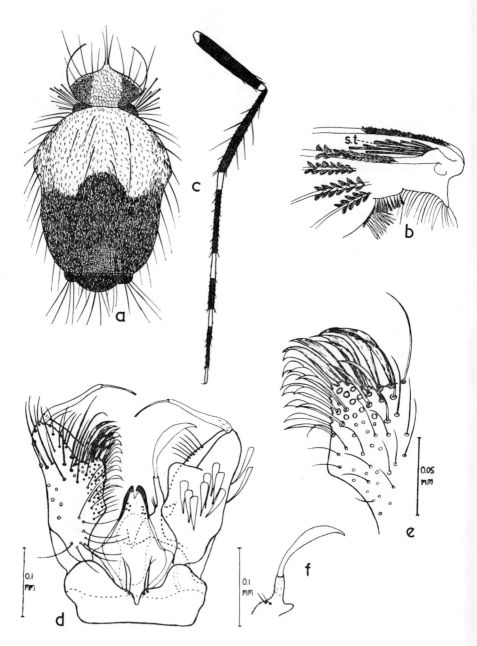

Fig. 60 *Aedes wattensis* Taylor. *a*, head and thorax; *b*, basal part of wing (*st*, scale tuft); *c*, hind leg; *d*, male terminalia; *e*, basal lobe; *f*, harpago. (*d-f*, after Marks.)

Style almost half length of coxite, pilose on basal sixth, slightly curved and tapering; appendage about three-fifths length of style. Harpago fifth length of coxite; appendage 2½ times length of harpago, curved and expanded on distal two-thirds with pointed tip. Paraproct with single tooth. Lobes of IXth tergite with 2 setae.

LARVA
Unknown.

BIOLOGY AND DISTRIBUTION
In Victoria, *Ae. wattensis* is known only from Pink Lakes, Underbool. It also occurs in Queensland, New South Wales, South Australia and Northern Territory.

Aedes (Chaetocruiomyia) macmillani Marks

Aedes macmillani Marks, 1964, *Proc. Linn. Soc., N.S.W.,* 89: 131-47.

Very similar to *Ae. wattensis* but much larger than *Ae. wattensis* (wing length 2·5-3·1 mm) and the posterior margin of the pale scutal area is rounded. The vertex has black scales anteriorly and pale ones posteriorly.

ADULT FEMALE
Vertex clothed with broad, flat, violet-black scales in front, becoming pale towards neck; upright scales dark. Palps and proboscis (fig. 61 e) black scaled. Scutum (fig. 61 a) clothed on anterior half with ochreous-white scales. Sometimes scutellum without flat scales. Posterior pronotum with dark narrow scales. Pleuron with large patches of white scales. No lower mesepimeral bristles. Wing black scaled with tuft of long narrow black scales at extreme base. Legs (fig. 61 c) black; all femora white at base; fore and mid femora with white knee patch. Fore and mid tarsi with narrow basal white bands on segments 1 and 2; hind tarsi with white basal bands on segments 1, 2 and 3, segment 5 white. Tergites black scaled with narrow basal white bands separated from large white lateral patches. Sternites I-V, white scaled with oblique lateral patches of black scales; VI, black with white lateral patches, white scales medially, and sometimes also along apical border; VII, black with small lateral patches of white scales.

ADULT MALE
Palps slightly longer than proboscis without labella. *Terminalia* (fig. 61 d). Similar to that of *Ae. wattensis* but differs as follows. On distal margin of basal lobe there are 2-4 long, strong setae, followed by 2-4 broad striated dark setae, and at apex of lobe and extending along distal third mesially are setae with curved, flattened tips. Appendage of harpago is broader. Lobes of IXth tergite with 3-5 setae.

LARVA
Unknown.

BIOLOGY AND DISTRIBUTION
A day-biting mosquito which has been collected in Victoria only during February. In Victoria it is a rare species: a few specimens have been

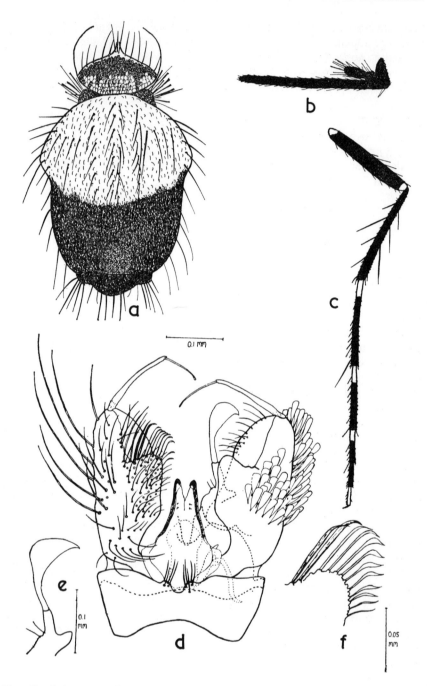

FIG. 61 *Aedes macmillani* Marks. *a*, head and thorax; *b*, proboscis and palp of female; *c*, hind leg; *d*, male terminalia; *e*, harpago; *f*, basal lobe. (*d-f*, after Marks.)

collected at Orbost, Cabbage Tree Creek, Wilson's Promontory, Sherbrooke and Lyonville. It has also been recorded from New South Wales and Flinders Island.

Subgenus PSEUDOSKUSEA Theobald
Pseudoskusea Theobald, 1907, *Mon. Cul.*, 4: 192.

Characters of the Subgenus
ADULT

Dark species without ornamentation. Proboscis longer than front femora. Palps of male about as long as proboscis; last two segments not swollen. Plume hairs of male antennae extending mainly dorsally and ventrally. *Male terminalia.* Coxite long, basal or apical lobe absent. Style long, slender, with long terminal appendage. Harpago absent. Phallosome simple and smooth.

LARVA

Antennae long. Head setae 5 and 6, long, single or 2 branched. Lateral comb a large patch of small scales. Siphon long, index not less than 4·0. Saddle covering about three-quarters of dorsal part of the anal segment. Anal papillae narrow, pointed, not less than half length of saddle.

The subgenus *Pseudoskusea* is restricted to the Australian Region and is represented in Victoria by three species.

Aedes (Pseudoskusea) bancroftianus Edwards
Aedes bancroftianus Edwards, 1921, *Bull. ent. Res.*, 12: 74.
Dobrotworsky, 1960d, 85: 257.

A rather small blackish mosquito. The head is clothed entirely with broad flat scales. The tarsi are unbanded; the claws of the fore and mid legs are toothed. The tergites have basal white bands.

ADULT FEMALE

Vertex (fig. 62 c) clothed with broad flat pale or brownish scales; a few upright scales pale or dark, in front and towards neck. Integument of thorax dark brown. Scutum clothed with narrow light-brown or dark-bronze scales, becoming creamy around front margin and bare area; there may be two lines of dark scales extending from scutellum for about half length of scutum. Posterior pronotum with elongate dark brown scales. Postspiracular area (fig. 62 d) with only a median patch of scales. Sternopleuron with a large patch of broad scales extending from below prealar area along posterior edge. Large patch of scales below upper mesepimeral bristles. Wings and legs dark scaled; hind femora pale except apical third or quarter and a dorsal line which are black. Fore and mid claws toothed, hind claws simple (fig. 62 e-g). Tergites II-V with complete basal creamy bands, VI with basal lateral creamy patches, sometimes forming basal bands. Sternites pale scaled with apical black bands on sternites IV-VI.

ADULT MALE

Palps about as long or longer than proboscis with labella (fig. 62 b). *Terminalia* (fig. 62 a). Coxite almost cylindrical with small dense patch of

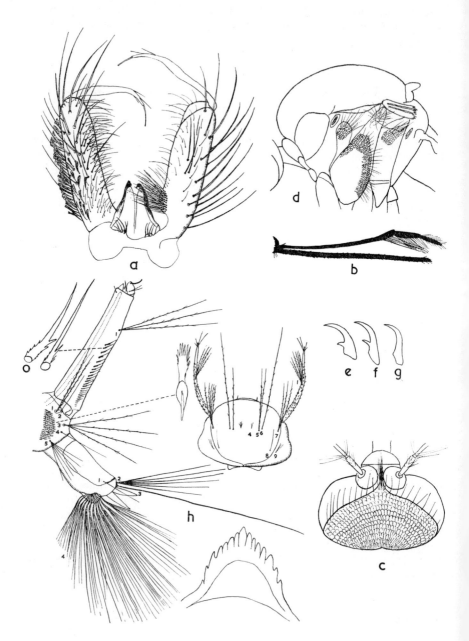

Fig. 62 *Aedes bancroftianus* Edwards. *a-g*, adult: *a*, male terminalia; *b*, proboscis and palp of male; *c*, female head; *d*, side view of thorax; *e-g*, female tarsal claws (*e*, fore; *f*, mid; *g*, hind). *h*, larva: head, mentum and terminal segments. (*a, e-h*, after Dobrotworsky.)

hairs at base. Basal and apical lobes absent. Style about three-fifths length of coxite with 3-4 long fine subterminal spines. Lobes of IXth tergite small with 5-6 setae.

LARVA (fig. 62 h)

Head and siphon pale. Head as long as broad. Antenna dark, curved, spiculate, as long as head; seta 1, 4-8 branched. Head setae: 4, 1-3 branched; 5 and 6, single; 7, 3-5 branched; 8, single; 9, 3 branched. Mentum with central large tooth and 7-8 lateral teeth on each side. Prothoracic setae: 1, 2, 4, 5 and 6, single; 3, 2 branched; 7, 3 branched. *Abdomen.* VIIIth segment: lateral comb patch of 70-80 fringed scales; seta 1, 2-3 branched; 2 and 4, single; 3, 4 branched; 5, 6 branched. Siphon tapering; index 4·0-4·5, mean 4·3; pecten of 21-23 spines; seta 1 arising two-thirds from base, long, usually 3 branched. Distance between base of distal spine of pecten and base of seta 1 less than width of siphon at level of seta 1. Seta 9 on ventral valve very stout, hook-like. Siphonal tracheae very narrow. Anal segment: saddle covering about three-quarters of dorsal part of segment; setae 1 and 3, single; 2, 8-9 branched; 4 (ventral brush) of 11-12 tufts. Anal papillae narrow, pointed, almost equal in length and about half length of saddle.

BIOLOGY

In Victoria larvae of *Ae. bancroftianus* have been found in fresh water pools in the Euroa district during the winter and spring months. This is open plain country just north of the Dividing Range, with sparsely scattered eucalypts, mainly along creeks and roads. In 1958 it was breeding in a string of pools in a natural watercourse with almost vertical banks. No larvae were found in these sites in 1959 but they were present in a roadside ditch 2-2½ feet deep, with grassy edges and reeds in some places; the water was cloudy.

The larvae behave like those of *Ae. theobaldi* (Taylor): most of the time they lie on their backs on the bottom of the pool, or attach themselves to the sides or to vegetation by means of strong hook-like setae on the spiracular valves. In contrast to the closely related *A. postspiraculosis, Ae. bancroftianus* is eurygamous. It is a day-biting mosquito which attacks man and domestic animals.

DISTRIBUTION

In Victoria *Ae. bancroftianus* occurs north of the Central Highlands. It is widely distributed in Australia.

Aedes (Pseudoskusea) postspiraculosis Dobrotworsky

Aedes postspiraculosis Dobrotworsky, 1960, *Proc. Linn. Soc. N.S.W.*, 85: 261.

Very similar to *Ae. bancroftianus* but the postspiracular area has two patches of scales. All the claws are simple.

ADULT FEMALE

Vertex clothed with broad flat scales, pale mesially and darker laterally; upright scales only towards neck. Proboscis and palps black scaled. Integument of thorax black. Scutum clothed with narrow brownish and yellowish-golden scales becoming paler and broader around bare area. Posterior pronotum with narrow and elongate brown scales, becoming broader and paler below. Postspiracular area (fig. 63 c) with patch of scales in middle and a second elongate patch on lower part between subspiracular area and sternopleuron. Large patches of broad scales also on sternopleuron and mesepimeron. Wings and legs dark; hind femora pale except at tip and along a dorsal line. All claws simple (fig. 63 e-g). Tergites (fig. 63 d) black scaled with creamy basal bands. Sternites creamy scaled with some admixture of black scales on segment VI; VII, black scaled.

ADULT MALE

Palps usually slightly shorter than proboscis with labella (fig. 63 b). *Terminalia* (fig. 63 a). Similar to that of *Ae. bancroftianus* but differs as follows: subterminal spines of style almost as thick as terminal appendage. Lobes of IXth tergite with 5-6 setae.

LARVA (fig. 63 h)

Milky white; head and siphon light brown, siphon becoming darker towards tip. Antenna dark, thin, curved, spiculate, as long as head; seta 1, 4-7 branched. Head setae: 4, 2-4 branched; 6 and 8, single; 5, single, rarely 2 branched on one side; 7, 4-5 branched, may be 3 branched on one side; 9, 2-5 branched. Mentum with long central tooth and 9-11 strong lateral ones on each side. Prothoracic setae: 1, 2, 3, 4, 5 and 6, single; 7, 3-4 branched. *Abdomen*. VIIIth segment: lateral comb patch of 60-70 fringed scales; seta 1, 3-4 branched; 2 and 4, single; 3, 4-5 branched; 5, 4-7 branched. Siphon tapering, with index 4·7-5·3, mean 5·0; pecten of 20-30 spines; seta 1 arising two-thirds from base, long, usually 3 branched, may be 4-5 branched. Distance between base of distal pecten spine and seta 1 greater than width of siphon at level of seta 1; seta 9 on ventral valve very stout, hook-like. Siphonal tracheae very narrow. Anal segment: saddle covering three-quarters of dorsal part of segment; setae 1 and 3, single; 2, 5-8 branched; 4 (ventral brush) of 11-12 tufts. Anal papillae narrow, almost equal in length, about half length of saddle.

BIOLOGY

Ae. postspiraculosis is confined to wooded undulating country. Natural watercourses which run only after heavy rains and retain water in holes for long periods provide the main breeding sites for this species. Water in such pools is usually cloudy. The pools may or may not have vegetation, but the banks are usually grassy; the depth varies from 1 to 2½ feet. The pools are usually shaded for part of the day and the water temperature remains below 68°F even during the summer. The larvae behave like those of *Ae. bancroftianus*, lying on their backs on the bottom for most of the time or attaching themselves to vegetation.

The number of generations depends on rainfall; in dried areas or during very dry summers there would be only spring and autumn generations, but in higher rainfall areas or during wet summers there may be in addition two or more summer generations.

Mating occurs during the day. The males form small swarms of a dozen or two near the breeding sites; they move about near the ground in

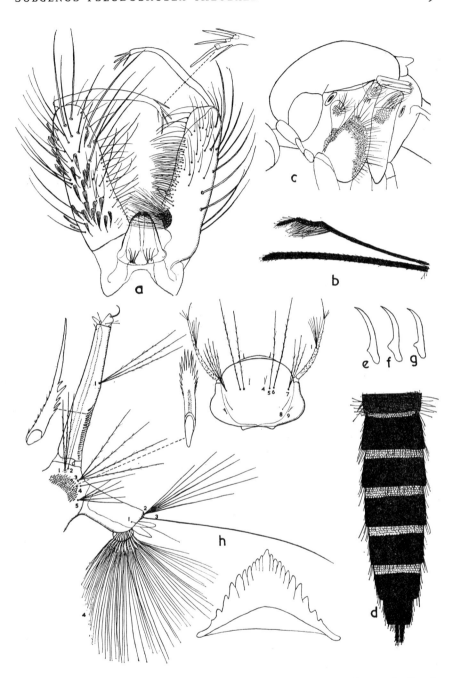

FIG. 63 *Aedes postspiraculosis* Dobrotworsky. *a-g*, adult: *a*, male terminalia; *b*, male proboscis and palp; *c*, side view of thorax; *d*, female abdomen, dorsal view; *e-g*, female tarsal claws (*e*, fore; *f*, mid; *g*, hind). *h*, larva: head, mentum, terminal segments. (*a, e-h*, after Dobrotworsky.)

'searching flights', and as females approach coupling occurs; it is usually completed on the grass. It is a stenogamous species. In the laboratory mating will take place in cages of 1 cubic foot capacity, if the mosquitoes have been induced to fly by shaking the cage or by just blowing into it. As soon as the females are in flight the males attack them and coupling can be observed.

Ae. postspiraculosis is a day-biting mosquito which attacks man, usually preferring to settle on clothing rather than on bare skin.

DISTRIBUTION

In Victoria it is widely distributed on and south of the Central Highlands. It has also been recorded from New South Wales and South Australia.

Aedes (Pseudoskusea) multiplex (Theobald)

Skusea multiplex Theobald, 1903, *Mon. Cul.,* 3: 293.
Pseudoskusea multiplex Theobald, 1907, *Mon. Cul.,* 4: 192.
Aedes multiplex Edwards, 1924, *Bull. ent. Res.,* 14: 386.
Dobrotworsky, 1960d, 85: 264.

A black mosquito with a transverse ochreous band across the middle of the scutum. The tergites may have bands but the tarsi are unbanded.

ADULT FEMALE

Narrow golden scales around eye margin. Vertex with broad flat, dark or pale, scales. Proboscis and palps dark scaled. Integument of thorax dark brown. Scutum (fig. 64 f) clothed with narrow curved bronze-black scales and with ochreous scales forming either broad transverse band across middle of scutum or two lateral patches. Posterior pronotum with a few narrow bronze scales. Postspiracular area (fig. 64 a) with bristles only. Sternopleuron with large patch of broad white scales below prealar area and a second patch along posterior edge; these two patches may join. Large patch of scales below upper mesepimeral bristles. Wings and legs dark scaled; hind femora pale on basal half, dark on apical half and with dark line dorsally. Fore and mid claws toothed, hind simple (fig. 64 c-e). Tergites black scaled, with basal lateral patches of white scales; in addition there may be a median basal patch of white scales on segments II and VI, and narrow basal bands on segments III-V. Sternites black scaled with white basal bands.

ADULT MALE

Palps about as long as proboscis with labella. Tergites with narrow basal bands; sternites white scaled with apical black bands. *Terminalia* (fig. 64 b). Coxite almost cylindrical, 4 times as long as wide with a small patch of moderately long, fine hairs at base; basal and apical lobes absent. Style long, almost straight, with 3 fine subapical spines. Lobes of IXth tergite with 2-6 setae.

LARVA (fig. 64 g)

Head and siphon brownish, body whitish. Head about three-quarters as long as broad. Antenna about four-fifths length of head; seta 1, 3 branched. Head setae: 4, 3-6 branched; 5 and 6, 2 branched; 7, 8-12 branched; 8, single; 9, 4-5 branched. Seta 6 of 2 unequal branches, inner branch much thinner and about three-quarters length of outer one. Mentum with longer central tooth and 11-12 lateral teeth on each side. Prothoracic setae: 1, 2, 3, 5 and 6, single; 3, may be 2 branched; 4, 2 branched; 7, 3 branched. *Abdomen.* VIIIth segment: lateral comb patch of about 60 fringed scales; seta 1, 6-7 branched; 2, 2 branched or single; 3, 6-7 branched; 4, single; 5, 6-7 branched. Siphon almost cylindrical, slightly tapering apically; index 4·1-4·6, mean 4·4; pecten of 18-24 spines; spines small with 1 or 2 teeth at base; seta 1 arising slightly beyond mid length, 2-3 branched, small. Anal segment: saddle covering about three-quarters of dorsal part of segment; setae 1 and 3, single; 2, 7-8 branched; 4 (ventral

FIG. 64 *Aedes multiplex* (Theobald). *a-f*, adult: *a*, side view of thorax; *b*, male terminalia; *c-e*, female tarsal claws (*c*, fore; *d*, mid; *e*, hind); *f*, thorax. *g*, larva: head, mentum and terminal segments. (*b-f*, *g*, after Dobrotworsky.)

brush) of 11-12 tufts. Anal papillae unequal, upper pair longer and about half length of saddle.

BIOLOGY

In Victoria *Ae. multiplex* is common only at Cabbage Tree Creek (East Gippsland). Larvae have not been collected in this state but in Queensland they have been found in fairly shaded ground pools (e.g. at Tewantin) 'in drying-out tea-tree swamps close to mangroves, water was fresh, discoloured, peaty soil, some dried grass on edge'. At Woombyne larvae have been found alone in a shallow well with earth walls and covered with wooden slabs (E. N. Marks). *Ae. multiplex* is a day-biting mosquito which attacks man.

DISTRIBUTION

In Victoria *Ae. multiplex* has been reported only from East Gippsland (Cabbage Tree Creek, Genoa, Kalimna) during January-April. It is a coastal species which is distributed from Queensland to East Gippsland.

Subgenus HALAEDES Belkin

Halaedes Belkin, 1962, *Mosquitoes of the South Pacific*, 1: 328. *Caenocephalus* Taylor, 1914, *Trans. Roy. ent. Soc. Lond.*, 1913: 700. *Aedes (Pseudoskusea)* in part of Edwards (1932: 157).

Characters of the Subgenus

ADULT

Decumbent scales on vertex narrow. Lower mesepimeral bristles present. Segment VIII of female abdomen partially retracted; cerci short and broad. Terminal segment of male palps swollen. *Male terminalia*. Basal and apical lobe of coxite absent. Harpago short, lobe-like, with numerous flattened, recurved setae.

LARVA

Antennae short. Lateral comb a very large patch of small scales. Siphon short, index 2·0-2·5; tracheae strongly swollen. Saddle reduced. Anal papillae very small, globular.

Aedes (Halaedes) australis (Erichson)

Culex australis Erichson, 1842, *Arch. Naturgesch.*, 8: 270.
Culex crucians Walker, 1856, *Ins. Saund. Dipt.*, 1: 432.
Culicada tasmaniensis Strickland, 1911, *Entom.*, 44: 181.
Caenocephalus concolor Taylor, 1914, *Trans. ent. Soc. Lond.*, 1913: 700. *Aedes concolor* Edwards, 1924, *Bull. ent. Res.*, 14: 387. *Aedes australis* Mattingly and Marks, 1955, *Proc. Linn. Soc. N.S.W.*, 80: 163.

The scutum is clothed with a mixture of golden and dark-bronze scales; the legs are dark, the tarsi unbanded. The tergites have broad creamy bands, usually constricted in the middle; the sternites are pale with apical lateral black patches.

Adult Female

Vertex (fig. 65 c) with narrow curved golden scales; upright scales black. Proboscis dark with some pale scales on basal half underneath. Scutum with mixture of golden and dark bronze scales. All scales on pleuron broad, elongate. Posterior pronotum dark scaled. Postspiracular area (fig. 65 d) with large patch of dark scales in middle and smaller patch of pale scales below on narrow part of area. Mixture of dark and pale scales on sternopleuron. Pale scales on mesepimeron; 4-6 lower bristles.

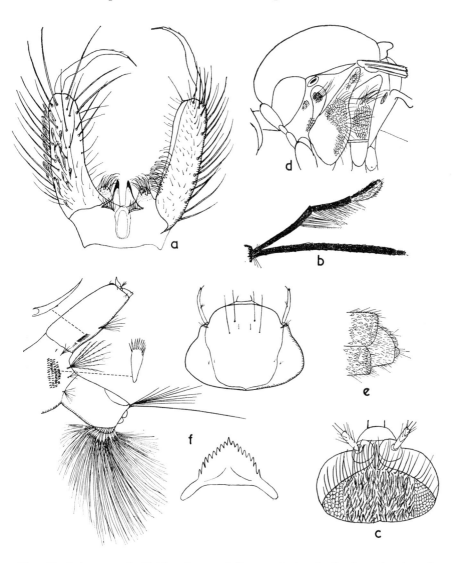

FIG. 65 *Aedes australis* (Erichson). *a-e*, adult: *a*, male terminalia; *b*, proboscis and palp of male; *c*, head; *d*, side view of thorax; *e*, terminal abdominal segments of female. *f*, larva: head, mentum and terminal segments.

Wings and legs dark scaled. All claw toothed. Tergites black scaled with basal creamy bands sometimes restricted in middle. Sternites pale scaled with apical lateral dark patches. Abdomen: VIIIth segment (fig. 65 e) rather large and only partly retractile; cerci short.

ADULT MALE

Palps (fig. 65 b) usually only slightly shorter than proboscis without labella but they may be only about four-fifths as long; terminal segment swollen. Fore and mid claws unequal, larger claws with two teeth, smaller with one; hind claws equal, toothed. *Terminalia* (fig. 65 a). Coxite long, narrow, with scales and long setae laterally, and small one tergally; on inner edge, towards base, numerous foliate setae. Basal lobe absent. Harpago short, lobe-like with numerous flattened, recurved setae. Style slender, curved, tapering, more or less swollen towards middle; 2-3 fine preapical setae; terminal appendage moderately long. Phallosome simple. Paraproct with single tooth. Lobes of IXth tergite with 6-7 setae.

LARVA (fig. 65 f)

Antenna and siphon short. Antennal seta 1 small, usually 2 branched. Head setae: 5, 6 and 8 usually single; 7, 2-4 branched. *Abdomen*. VIIIth segment: lateral comb patch of more than one hundred scales; setae 1, 2, 4 and 5, 2-3 branched; 3, 6-8 branched. Siphon index about 2: seta 1 moderately long, 7-8 branched, arising at about mid length; pecten of 10-15 spines, 10 of which are closely set. Anal segment: seta 1 very small tuft; 2, 8-10 branched; 3, single; 4, about 15 tufts. Saddle reduced. Anal papillae very small, globular.

BIOLOGY

Ae. australis breeds in salt water; larvae can be found in rock and ground pools just above high tide level on the coast. It is one of the few mosquitoes which combine stenogamy with autogeny.

DISTRIBUTION

It is widely distributed along the south-east coast of Australia from the southern border of Queensland to South Australia and the southern part of Western Australia. It is also recorded from Bass Strait islands, Tasmania, Lord Howe and Norfolk Islands.

Genus CULISETA Felt

Theobaldia Neveu-Lemaire, 1902, *C. R. Soc. Biol. Paris*, 54: 1331 (*non* Fisher 1885). *Culiseta* Felt, 1904, *Bull. N.Y. St. Mus.* No. 79: 391c. *Theobaldinella* Blanchard, 1905, *Les Moustiques:* 390. *Pseudotheobaldia* Theobald, 1907, *Mon. Cul.*, 4: 271.

This is a small genus of 38 species most of which are restricted to cooler areas: fourteen occur in the Palearctic, ten in the Nearctic and twelve in the Australian Region. Only two species have been recorded in the Ethiopian Region, three in India and two in Mexico. In Australia most of the species occur in the eastern states and the number decreases from south to north. Of the eleven species, one occurs in Queensland, two in New South Wales and nine in Victoria. Three are recorded in Tasmania, two in South Australia and one in the southern part of Western Australia. A single species is known from New Zealand.

Characters of the Genus

ADULT

Head with narrow, curved and upright forked scales on vertex. Proboscis moderately long. Palps of male about as long as proboscis; in female palps short. Spiracular bristles (fig. 4 A) always present but fine and few in number in Australian species. Postspiracular area usually bare (a few minute scales present in *C. littleri*). Lower mesepimeral bristles present. Female claws simple. Pulvilli absent. Subcosta (fig. 5C) with group of seta on underside at base. Abdomen of female bluntly rounded, with non-retractile VIIIth segment; in Australian species tergites dark scaled. *Male terminalia*, Coxite rather long; basal lobe present; apical lobe absent in Australian species. Style simple, long and slender, with terminal spine. Paraproct with a few terminal teeth.

LARVA

Siphon rather long with a pair of siphonal tufts or a pair of setae at its base (except *C. littleri* which has single or two branched pair of setae half way along siphon). The adults of Culiseta do not provide satisfactory characters for subgeneric division and the subgenera recognized by Edwards (1932), as well as those established more recently, are based on larval morphology. Four subgenera, two exclusively Australian, are represented in Victoria: *Culicella* with five species, *Neotheobaldia* with two and *Climacura* and *Austrotheobaldia* each with one.

KEYS TO VICTORIAN SPECIES OF THE GENUS CULISETA

Adult Females

1   Proboscis pale scaled beneath .. .. *inconspicua*
      Proboscis entirely black 2
2(1) Tarsi entirely dark 3
      Last 2-3 segments of all tarsi pale 5
3(2) Upright scales on vertex black .. .. *antipodea*
      Upright scales on vertex pale 4
4(3) Mesepimeron with patch of flat scales towards middle .. .. *littleri*
      Mesepimeron with patch of hairs towards middle .. .. *otwayensis*
5(2) 2-3 basal segments of tarsi without pale scales at base 6
      2-3 basal segments of tarsi with pale scales 8
6(5) Upright scales on vertex mostly pale; smaller species .. .. .. *hilli*
      Upright scales on vertex dark 7
7(6) Upright scales on vertex dark brown; integument reddish; larger
      species .. .. *frenchii*
      Upright scales on vertex black; integument brown; smaller species
      .. .. *sylvanensis*

8(5) Upright scales on vertex mostly pale; mesepimeron with patch of hairs towards middle .. .. *victoriensis*
Upright scales on vertex black; mesepimeron with patch of elongate scales towards middle .. .. *drummondi*

Larvae (fourth-stage)

1   Pair of tufts at base of siphon 2
No tufts at base of siphon, a single pair of tufts half-way along siphon .. .. *littleri*
2(1) Siphon with a pair of basal tufts and row of more than ten ventral tufts; comb teeth in a single regular row .. .. *antipodea*
Siphon without ventral tufts; comb teeth a large patch of scales .. 3
3(2) Pecten consists of a row of hairs 4
Pecten consists of spine-like teeth 5
4(3) Siphonal tuft 2 branched .. .. *hilli*
Siphonal tuft a single hair .. .. *frenchii*
5(3) Lateral comb consists of stout, short tooth-like scales .. *drummondi*
Lateral comb consists of fringed scales 6
6(5) Lava brown 7
Larva milky-white 8
7(6) Pecten consists of 3-5 spines; head seta 5, 3-5 branched, about half length of seta 6 .. .. *otwayensis*
Pecten consists of 9-10 spines; head seta 5-2 branched, about four-fifths of length of seta 6 .. .. *inconspicua*
8(6) Head seta 6, long, single or 2 branched; seta 5, 3-5 branched, about three-quarters length of seta 6 .. .. *victoriensis*
Head seta 6 of moderate length, 3-5 branched; seta 5, 4-7 branched, nearly as long as seta 6 .. .. *sylvanensis*

Subgenus AUSTROTHEOBALDIA Dobrotworsky

Austrotheobaldia Dobrotworsky, 1954, *Proc. Linn. Soc. N.S.W.*, 79: 68.

Characters of the Subgenus

LARVA

All hairs simple. Head large and broad. Antenna long; two apical bristles removed from tip. Siphon long, tapering; seta 1 half-way along siphon, single or 2 branched; pecten of triangular teeth. Lateral comb a row of small scales and two or three irregular short rows of long scales. Seta 2 of anal segment branched. Anal papillae about as long as saddle, pointed.

Culiseta (Austrotheobaldia) littleri (Taylor)

Chrysoconops littleri Taylor, 1914, *Trans. ent. Soc. Lond.*, (4): 702. *Theobaldia littleri* Edwards, 1924, *Bull. ent. Res.*, 14: 363. *Culiseta littleri* Stone, Knight and Starke, 1959, *Syn. Cat. Mosquitoes World*, 6: 220. Dobrotworsky, 1954, 79: 75.

The upright scales on the vertex are pale. The tarsi are entirely dark.

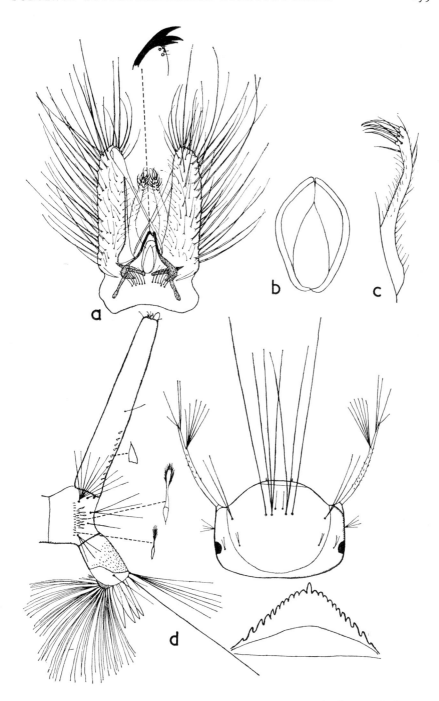

FIG. 66 *Culiseta littleri* (Taylor). *a-c*, adult: *a*, male terminalia; *b*, phallosome; *c*, basal lobe of coxite. *d*, larva: head, mentum and terminal segments. (*a, d*, after Dobrotworsky.)

Adult Female

Upright scales on vertex pale. Palps and proboscis brown. Integument brown. Scutum clothed with pale golden scales. Posterior pronotum with narrow curved scales. Two to three fine, pale spiracular bristles. Postspiracular area with a few minute scales. Mesepimeron with patch of scales towards middle and usually 3 lower bristles. Tarsi entirely dark brown. Venter pale with light brown reflections.

Adult Male

Palps slightly shorter than proboscis with labella. Venter light brown, sometimes with lateral triangular areas of brown and black scales. *Terminalia* (fig. 66 a-c). Coxite almost three times as long as broad. Basal lobes about two-thirds of length of coxite, separated for nearly its whole length and with rather strong spines on tip. Paraproct with 3 teeth. Lobe of IXth tergite with 4-7 setae.

Larva (fig. 66 d)

Reddish brown; head light brown, siphon brown. All setae non-plumose. Head broader than long. Antenna long, thin and curved; seta 1, 7-8 branched. Head setae: 4 and 6, single; 5 and 7, 2 branched; 8 and 9, 2-3 branched. Prothoracic setae: 1, 2, 5 and 6, single; 3, 4 and 7, 2 branched. *Abdomen*. VIIIth segment: lateral comb of 16-30 small scales in basal row and 8-15 long scales in 2-3 short distal rows; seta 1, 7-9 branched; 2 and 4, single; 3, 3-5 branched; 5, 3 branched. Siphon with index about 7; seta 1, single or 2 branched, about half way along siphon; basal tuft absent; pecten of 13-15 triangular teeth. Anal segment: saddle complete ring; seta 1, single; 2, 5-6 branched; 3, 3 branched (1 long and 2 short); 4, of 10-11 tufts, 2 small, precratal. Anal papillae, narrow, pointed, about as long as saddle.

Biology

C. littleri breeds in highland forests. The larvae are found in heavily shaded permanent pools or sometimes in pools under logs or uprooted trees. In all those situations the water is clean and cold. *C. littleri* occasionally attacks man.

Distribution

Restricted to the Eastern Highlands and the Otway Ranges of Victoria where it breeds in areas with an average annual rainfall of 50 inches. It has also been recorded from New South Wales and Tasmania.

Subgenus CULICELLA Felt

Culicella Felt, 1904, *N.Y. St. Mus. Bull.*, 79: 391.

Characters of the Subgenus

Larva

Antennae long with large tuft well beyond middle. Mouth brushes very large. Siphon long, tapering, with one pair of basal tufts; pecten consisting of spine-like teeth only, no fine hairs distally. Lateral comb a large patch of scales.

Culiseta (Culicella) victoriensis (Dobrotworsky)

Theobaldia victoriensis Dobrotworsky, 1954, *Proc. Linn. Soc. N.S.W.*, 59: 73. *Culiseta victoriensis* Stone, Knight and Starke, 1959, *Syn. Cat. Mosquitoes World,* 6: 221. Dobrotworsky, 1954, 79: 73.

This species can be recognized by the pale upright scales on the vertex; the first three tarsal segments are banded, the fourth and fifth pale.

ADULT FEMALE

Forked upright scales on vertex mostly pale. Proboscis and palps black scaled. Integument brown. Scutum clothed with pale golden scales. Posterior pronotum with narrow curved scales. Three pale spiracular bristles. Mesepimeron with patches of hairs towards middle and 2-4 lower mesepimeral bristles. Knee spots conspicuous. First three tarsal segments of all legs with pale scales at base forming narrow ring (fig. 67 d); fourth and fifth segments entirely pale. Sternites pale scaled with median patches of black scales.

ADULT MALE

Palps (fig. 67 c) as long as proboscis with labella. Venter pale scaled with large median patches of black scales, or black scaled with pale scales apically on sternites. *Terminalia* (fig. 67 a, b). Coxite more than twice as long as broad with dark scales and long bristles apically and laterally; tergally coxite bears numerous long, goldish, mesially directed setae; basal lobe about two-thirds of length of coxite and separated from it for about half length. Paraproct with 4 large teeth. Phallosome simple, widening distally with several small teeth at top. Lobes of IXth tergite with patch of 10-14 setae.

LARVA (fig. 67 e)

Milky-white with black setae. Head large, pale. Antenna long, thin and curved; seta 1 of 14-16 slightly plumose branches arising about three-quarters of length from base. Head setae; 4, single or 2 branched; 5, 3-5 branched; 6, single or 2 branched, long; 7, 6-7 branched; 8 and 9, 2 branched. Prothoracic setae: 1, 2, 5 and 6, single; 3 and 4, 2 branched; 7, 3 branched. *Abdomen.* VIIIth segment: lateral comb large triangular patch of fringed scales; seta 1, 4-5 branched; 2 and 4, single; 3, 2-3 branched; 5, 2 branched. Siphon long, gradually tapering, with index 6·3-6·8; basal siphonal seta usually 2 branched; pecten of 10-13 spine-like teeth. Anal segment: saddle, complete ring; seta 1, 2-5 branched; 2, 7-8 branched; 3, single; 4 (ventral brush) of 15 tufts. Anal papillae narrow, pointed, more than twice as long as saddle.

BIOLOGY

C. victoriensis breeds in subterranean water in tunnels of land cray-fish (*Engaeus* spp.). It overwinters in the larval stage and the spring generation usually appears in October. Adults become numerous during the early summer and remain common until early autumn. In the laboratory eggs have been laid singly on soil.

C. victoriensis is a man-biting species which is very troublesome in forests.

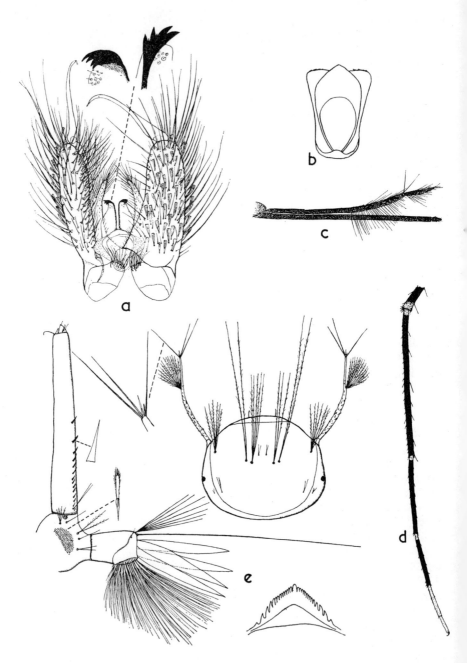

Fig. 67 *Culiseta victoriensis* (Dobrotworsky). *a-d*, adult: *a*, male terminalia; *b*, phallosome; *c*, proboscis and palp of male; *d*, hind tarsus. *e*, larva: head, mentum and terminal segments. (*a*, *e*, after Dobrotworsky.)

DISTRIBUTION

C. victoriensis is distributed in the Eastern Highlands, South Gippsland uplands and Wilson's Promontory. It is more numerous in areas with an average annual rainfall above 35 inches. It is not, as yet, recorded from other states.

Culiseta (Culicella) drummondi (Dobrotworsky)

Theobaldia drummondi Dobrotworsky, 1960, *Proc. Lin.. Soc. N.S,W.*, 85: 241. *Culiseta drummondi* Stone, 1961, *Proc. ent. Soc. Wash.*, 63: 44.

Very similar to *C. victoriensis* but the upright scales on the vertex are black, and usually the first two tarsal segments are banded, the last three pale.

ADULT FEMALE

Upright forked scales on vertex black. Palps short, integument light brown. Scutum clothed with narrow curved golden and dark-bronze scales. Posterior pronotum with some narrow pale scales and hairs. Three spiracular bristles. Mesepimeron with patch of narrow pale scales towards middle; 3-6 lower mesepimeral bristles. Tarsi with narrow basal bands on segments 1-2, segments 3-5 pale. All claws simple. Wing membrane with a faint blotch near middle. Sternites creamy, with some admixture of dark scales, or dark scaled with admixture of light ochreous scales.

ADULT MALE

Palps as long as proboscis with labella. Fore and mid claws toothed, hind claws simple. Sternites usually dark scaled with admixture of pale scales increasing towards end of abdomen. *Terminalia* (fig. 68 a, b). Coxite more than twice as long as broad with dark scales sternally and laterally. On sternal aspect coxite bears long strong setae laterally and apically. Tergally, coxite has numerous long goldish mesially directed setae. Basal lobe of coxite about two-thirds length of coxite, with tuft of long curved setae on tip. Paraproct with 6 teeth. Phallosome simple, widening distally, with several small teeth on top. Lobes of IXth tergite with 12-14 long strong setae.

LARVA (fig. 68 c)

Head yellowish; body milky-white; siphon brown; thoracic setae black. Antenna curved, slightly shorter than length of head; seta 1 with 19-24 branches. Head setae: 4, small, 2 branched; 5, single or 2-3 branched; 6, single; 7, 3-7 branched; 8 and 9, 2-4 branched. Mentum with long central tooth and 13 lateral teeth on each side. Prothoracic setae single, except seta 4, which is a minute tuft, and seta 7, which is 2-3 branched. *Abdomen*. VIIIth segment: lateral comb triangular patch of more than a hundred stout tooth-like scales; seta 1-5, small tufts. Siphon very long, gradually tapering, with index 8·1-9·3, mean 8·6; basal siphonal seta, a small tuft; pecten of 8-11 flattened spines. Anal segment: saddle, complete ring; seta 1, small tuft; 2, 6-7 branched; 3, single; 4 (ventral brush) of 13-14 tufts. Anal papillae narrow, pointed, almost twice as long as saddle.

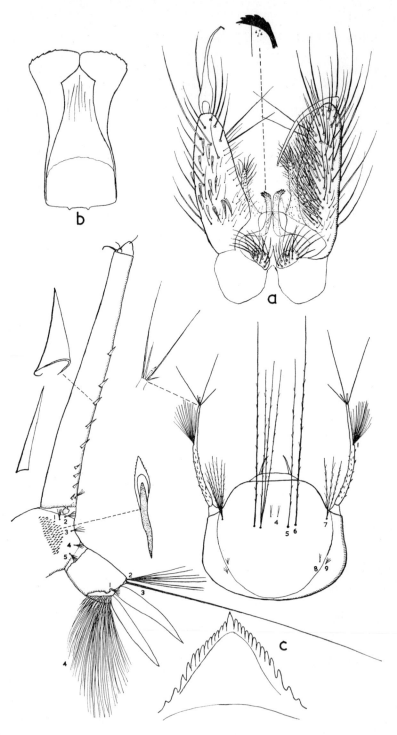

FIG. 68 *Culiseta drummondi* (Dobrotworsky). *a-b*, adult: *a*, male terminalia; *b*, phallosome. *c*, larva: head, mentum and terminal segments. (after Dobrotworsky.)

Biology

C. drummondi was found on a west-facing hillside with a good cover of forest and undergrowth, breeding in pits some 2-3 feet deep, concealed by undergrowth, fallen branches and other debris. The water in them was more or less cloudy and its temperature remained low even during the summer. The larvae and pupae are able to develop normally only at low temperatures. In the laboratory they produced adults successfully at 56-59°F. Temperatures above 68°F were fatal for larvae and pupae: at 74°F all larvae died within six days and pupae died without producing adults.

This is a day-biting species which attacks man freely near its breeding places.

Distribution

C. drummondi is known only from one locality, Sylvan, Victoria.

Culiseta (Culicella) sylvanensis (Dobrotworsky)

Theobaldia sylvanensis Dobrotworsky, 1960, *Proc. Linn. Soc. N.S.W.*, 85: 245. *Culiseta sylvanensis* Stone, 1961, *Proc. ent. Soc. Wash.*, 63: 44.

Usually all the upright scales on the vertex are black but there may be some admixture of pale ones. The tarsi are unbanded; the last two or three segments are pale.

Adult Female

Vertex with all upright scales black or with some admixture of paler scales. Palps short. Integument brown. Scutum clothed mainly with dark-bronze narrow curved scales. Posterior pronotum with narrow curved scales. One or two very small spiracular bristles. Mesepimeron with 2-3 lower bristles and patch of hairs and narrow scales towards middle. Legs black. Last 2-3 segments of tarsi pale. Claws simple. Venter yellowish scaled.

Adult Male

Palps longer than proboscis without labella. Claws of fore and mid legs toothed, hind simple. *Terminalia* (fig. 69 a, b). Coxite about twice as long as broad, with black scales sternally and laterally. Coxite with long, strong, black setae laterally and distally on tergal aspect, short dense setae on mesial aspect. Basal lobe about two-thirds length of coxite, with tuft of long curved setae. Paraproct with five teeth. Phallosome simple, only slightly widened distally, with small teeth at top. Lobes of IXth tergite with 8-10 long setae.

Larva (fig. 69 c)

Head and siphon yellowish; body milky-white; setae black. Antenna about two-thirds length of head; seta 1, 5-10 branched. Head setae: 4, 3-5 branched; 5, 4-7 branched; 6, 3-5 branched; 7, 4-8 branched; 8, single; 9, 2-5 branched. Setae 5, 6 and 7 plumose. Mentum with 11-12 lateral teeth on each side. Prothoracic setae single except setae 7, which is 2 branched, and seta 3, which may be 2 branched. *Abdomen*, VIIIth segment: lateral comb patch of more than a hundred fringed scales; seta 1, 2-4 branched; 2 and 4, single; 3, 2-4 branched; 5, 2-3 branched. Setae

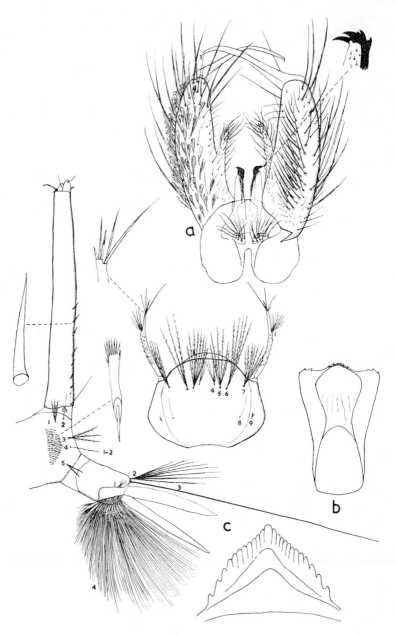

FIG. 69 *Culiseta sylvanensis* (Dobrotworsky). *a-b*, adult: *a*, male terminalia; *b*, phallosome. *c*, larva: head, mentum and terminal segments. (after Dobrotworsky.)

1 and 3 plumose. Siphon long, slightly tapering, with index 5·5-7·9, mean 6·7; basal seta long, single; pecten of 8-11 spines. Anal segment: saddle complete ring; seta 1, small tuft; 2, 4-6 branched; 3, single; 4 (ventral brush) of 14 tufts. Anal papillae narrow, pointed, about twice as long as saddle.

BIOLOGY AND DISTRIBUTION
Similar to those of *C. drummondi*.

Culiseta (Culicella) otwayensis (Dobrotworsky)

Theobaldia otwayensis Dobrotworsky, 1960, *Proc. Linn. Soc. N.S.W.*, 85: 346. *Culiseta otwayensis* Stone, 1961, *Proc. ent. Soc. Wash.*, 63: 44.

Very similar to *C. littleri* but in place of the broad scales towards the middle of the mesepimeron there is a patch of hairs and narrow scales.

ADULT FEMALE
Upright scales on vertex creamy. Proboscis black scaled. Palps short, integument light brown. Scutum clothed with dark-brown narrow curved scales. Posterior pronotum with hairs and narrow scales. Two to three small spiracular bristles. Mesepimeron with 1-2 lower bristles; some hairs towards middle. Legs dark. Venter clothed with brownish scales.

ADULT MALE
Palps slightly shorter than proboscis with labella. *Terminalia* (fig. 70 a, b). Coxite more than twice as long as broad, with long, strong setae and fine ones. Basal lobe about two-thirds length of coxite, with a tuft of rather long, curved, thick setae. Style as long as coxite, almost straight. Phallosome smooth, simple, oval in shape. Paraproct with 3 teeth. Lobes of IXth tergite inconspicuous, with 1-2 strong setae.

LARVA (fig. 70 c)
Brownish. Head about three-fifths as long as broad. Antenna slightly shorter than length of head; seta 1 about 2-5 branched. Head setae: 4 moderately long, single or 2 branched; 5, 3-5 branched; 6, single; 7, 3-4 branched; 8 and 9, 2-3 branched. Mentum small, with large central tooth and 13-14 lateral teeth. All prothoracic setae single but seta 4 may be 2 branched on one side. *Abdomen*. VIIIth segment: lateral comb patch of 50-60 fringed scales; seta 1, plumose, 8-9 branched; 2 and 4, single; 3, plumose, 5-6 branched; 5, 3-4 branched. Siphon slightly tapering; index 4·8-5·6; basal seta single; pecten of 3-5 spines. Anal segment: saddle complete ring; seta 1, 2 branched, about one-thirds of length of saddle; 2, 6-8 branched; 3, with 1 long and 2 short branches; 4 (ventral brush) of 15-17 tufts. Anal papillae narrow, pointed, about as long as saddle.

BIOLOGY AND DISTRIBUTION
C. otwayensis is known from only a single locality in the Otway Ranges where larvae were found in a pool, under an uprooted tree and shaded for most of the day. The water was slightly clouded, with a temperature of 58°F. No adults have been found in the field and nothing is known about their biting habits.

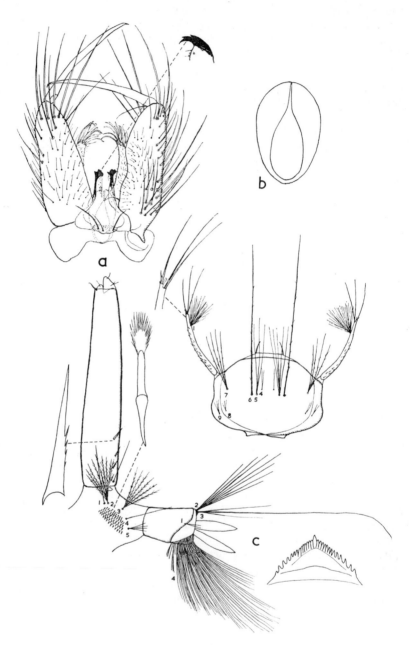

Fig. 70 *Culiseta otwayensis* (Dobrotworsky). *a-b*, adult: *a*, male terminalia; *b*, phallosome. *c*, larva: head, mentum and terminal segments. (after Dobrotworsky.)

Culiseta (Culicella) inconspicua (Lee)

Theobaldia inconspicua, Lee, 1937, *Proc. Linn. Soc. N.S.W.,* 42: 294-298. *Culiseta inconspicua* Stone, Knight and Starcke, 1959, *Syn. Cat. Mosquitoes World,* 6: 220. Dobrotworsky, 1954, 79: 75.

Distinguished from all other species of the genus in having the proboscis pale on the underside.

Adult Female

Upright scales on vertex dark brown, sometimes pale in front. Palps short. Proboscis dark brown above, pale beneath. Integument light brown. Scutum clothed with narrow curved dark-brown scales. Posterior pronotum bare. Two or three small spiracular bristles. Mesepimeron with a patch of hairs and elongate scales towards middle; 2-3 lower bristles. Tarsi entirely dark, claws simple. Venter clothed with dark scales basally, becoming pale apically.

Adult Male

Palps (fig. 71 c) as long as proboscis with labella, penultimate and terminal segments turned upwards. *Terminalia* (fig. 71 a, b). Coxite more than twice as long as broad; basal lobe very small and imperfectly separated from coxite. Phallosome smooth, simple, oval in shape. Paraproct with 3 teeth. Lobes of IXth tergite with 6-9 long setae.

Larva (fig. 71 d)

Brown. Head large. Antenna long; seta 1 arising about four-fifths of length from base, 18-23 branched. Head setae: 4 and 6, single; 5, 8 and 9, 2 branched; 7, 4 branched. Prothoracic setae: 1, 2, 5 and 6, single; 3, 4 and 7, 2 branched. *Abdomen.* VIIIth segment: lateral comb large patch of about 60 fringed scales; seta 1, 6-7 branched; 2 and 4, single; 3, 5-8 branched; 5, 3-7 branched. Siphon cylindrical, slightly broader towards mid length, index $5 \cdot 0$-$5 \cdot 5$; basal seta single, long; pecten of 9-14 spines. Anal segment: saddle complete ring; seta 1, single; 2, 7-10 branched; 3, 3 branched, two short, one long; 4 (ventral brush) of 13-14 tufts. Anal papillae narrow, pointed, almost equal and about two-thirds length of saddle.

Biology

C. inconspicua is a homodynamous species which breeds in a large variety of shaded ground and rock pools and swamps overgrown by reeds. The pools may be shallow or deep; the water may be clean or be polluted by decaying leaves and debris. The temperature in such pools is usually 58-65°F. Eggs are laid on water surface in bowl-shaped raft.

C. inconspicua does not attack man, and the blood sources of this species have not been established.

Distribution

It is the most widely distributed of all *Culiseta* species; in Victoria it is usually confined to woodlands approximately south of the average annual isohyet of 20 inches. The species is also recorded from New South Wales, South Australia and Tasmania.

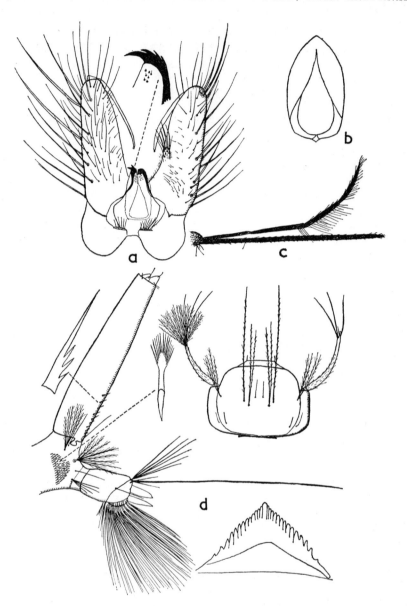

Fig. 71 *Culiseta inconspicua* (Lee). *a-c*, adult: *a*, male terminalia; *b*, phallosome; *c*, male proboscis and palp. *d*, larva: head, mentum and terminal segments.

Subgenus CLIMACURA Howard, Dyar and Knab

Climacura Howard, Dyar and Knab, 1915, *Mosq. N. and C. Amer.*, 3: 452.

Characters of the Subgenus

Head very large. Antennae long with large tuft. Mouth brushes very large.

Siphon long, tapering, with one pair of small basal tufts and more than ten ventral tufts, extending nearly to the end of the siphon. Comb teeth in a single regular row.

Culiseta (Climacura) antipodea Dobrotworsky
Culiseta antipodea Dobrotworsky, 1962, *Proc. Linn. Soc. N.S.W.*, 67: 291.

A very dark, almost black, mosquito. The upright scales on the vertex and the scales on the scutum, tergites and sternites are black. The tarsi are also entirely black.

ADULT FEMALE
Vertex clothed with narrow curved pale scales and black forked scales. Proboscis black scaled above with pale reflections below. Scutum sparsely clothed with narrow curved black scales. Scutal bristles black. Posterior pronotum with a few hair-like scales. Two or three spiracular bristles. One to three lower mesepimeral bristles and a patch of pale hairs towards the middle. Legs black scaled. Tergites and sternites black scaled.

ADULT MALE
Palps about as long as proboscis with labella. Antennal hairs of verticils evenly spread around segments. *Terminalia* (fig. 72 a, b). Coxite about three times as long as broad, with black bristles. Basal lobe about one-fifth of length of coxite, with 2-3 long strong setae and several finer ones at tip. Style narrow, slightly less than half length of the coxite. Terminal appendage small. Paraproct with four strong teeth. Phallosome simple, smooth. Lobes of IXth tergite flat, with 5-6 setae.

LARVA (fig. 72 c, d)
Brown. Head broad. Antenna long, curved, about as long as head; seta 1, with about 30 branches. Head setae: 4 and 6, single; 5, 4-5 branched, about quarter length of seta 6; 7, 6-8 branched; 8, 3-5 branched; 9, 4-6 branched. Mentum with broad central tooth and 6-7 lateral teeth on each side. Prothoracic setae: 1, 2, 5 and 6, single; 3, 2 branched; 4, 4-7 branched; 7, 3 branched. *Abdomen*. VIIIth segment: lateral comb of 14-21 scales in single row; seta 1, 5-9 branched; 2 and 4, single; 3, 6-9 branched; 5, 4-5 branched. Siphon long, slightly tapering, index 6·1-7·5, mean 6·8; pecten of 6-9 spines; basal siphonal setae, small, 3-5 branched; 7-8 minute lateral setae along siphon; 10-12 minute setae on ventral side between pectens, and a row of 5-6 single, or 2 branched, longer, fine setae in single row above pecten. Anal segment: saddle complete ring; seta 1, small, 3-4 branched; 2, 6-10 branched; 3, single; 4 (ventral brush) of 14 tufts. Anal papillae narrow, pointed, about as long as saddle.

BIOLOGY
C. antipodea breeds in tea tree swamps and semi-permanent pools 2-3 feet deep, in coastal heath country. In a swamp near Cann River the vegetation was so dense and entangled that movement through it was possible only by following animal tracks. Eggs are laid on the water surface in rafts. *C. antipodea* does not attack man and nothing is yet known about the blood sources of this species.

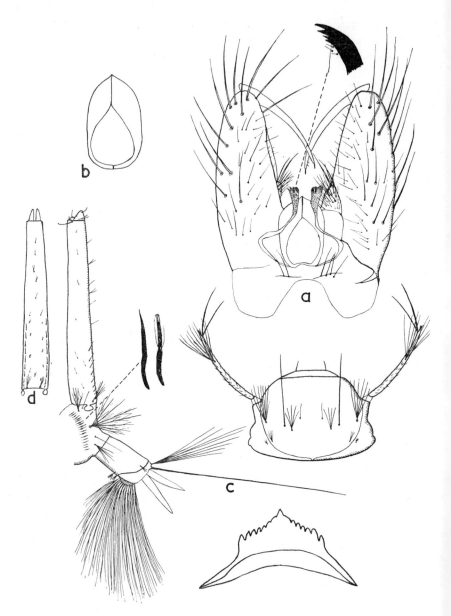

Fig. 72 *Culiseta antipodea* Dobrotworsky. *a-b*, adult: *a*, male terminalia; *b*, phallosome. *c-d*, larva: *c*, head, mentum and terminal segments; *d*, siphon, ventral view. (after Dobrotworsky.)

DISTRIBUTION

C. antipodea is known from East Gippsland (Cann River) and Wilson's Promontory in Victoria, and from south Queensland.

Subgenus NEOTHEOBALDIA Dobrotworsky

Neotheobaldia Dobrotworsky, 1954, *Proc. Linn. Soc. N.S.W.*, 79: 68. *Neotheobaldia* Dobrotworsky, 1958, *Proc. ent. Soc. Wash.*, 60: 186.

Characters of the Subgenus

LARVA

Head not very large. Antennae of moderate length; seta 1 well beyond middle; none of apical setae removed from tip. Siphon of moderate length, with one pair of basal tufts; pecten teeth in form of hairs. Saddle, complete ring; no precratal tufts. Seta 3 of anal segment single. Anal papillae not large, only slightly longer than saddle.

Culiseta (Neotheobaldia) hilli (Edwards)

Theobaldia hilli Edwards, 1926, *Bull. ent. Res.*, 17: 111.
Culiseta hilli Stone, Knight and Starke, 1959, *Syn. Cat. Mosquitoes World*, 6: 222. Dobrotworsky, 1954, 79: 69.

Similar to *C. frenchii* but the upright scales on the vertex are mostly pale. The venter has more or less conspicuous median patches of black scales.

ADULT FEMALE

Upright scales on vertex mostly pale, becoming dark towards neck. Proboscis and palps clothed with blackish scales. Integument brown. Scutum clothed with pale golden narrow curved scales. Posterior pronotum with narrow curved scales. Two to three small pale spiracular bristles. Mesepimeron with patch of scales and hairs towards middle, 2-3 lower bristles. Last 2-3 segments of tarsi pale. Venter pale scaled, sometimes with median patches of black scales.

ADULT MALE

Palps slightly shorter than proboscis with labella. Venter usually with sternites I-IV entirely black scaled, and sternites V-VII more or less pale apically; venter may be pale with median patches of black scales. Terminalia (fig. 73 a, b). Coxite short, about twice as long as broad. Basal lobe large, well separated, reaching nearly to tip of coxite and bearing long curved setae at tip. Paraproct with 4 strong teeth and 4 shorter ones. Phallosome simple, widening distally, with small teeth on tip. Lobes of IXth tergite with 9-15 fine setae.

LARVA (fig. 73 c)

Milky-white with black setae. Head and siphon light brown, siphon darker towards end. Antenna slightly shorter than head; seta 1 arising two-thirds of length from base, 7-14 branched. Head setae: 4, 3-4 branched; 5, 3-6 branched; 6, 3-4 branched; 7, 5-8 branched; 8, single; 9, 2-4 branched. Prothoracic setae: 1, 2, 4, 5 and 6, single; 3 and 7, 2 branched. *Abdomen.* VIIIth segment: lateral comb large patch of fringed scales; seta 1, 3 branched; 2 and 4, single; 3, 2-4 branched; 5, 2-3 branched. Siphon cylindrical, tapering distally, index 5·0-5·4; basal seta 2 branched; pecten teeth replaced by 8-10 hairs. Anal segment: saddle, complete ring; seta

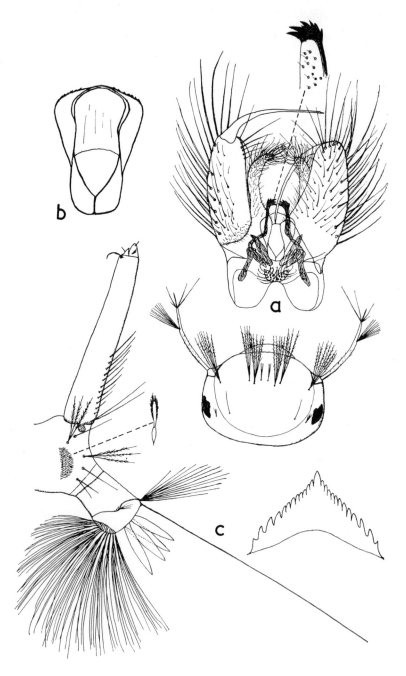

Fig. 73 *Culiseta hilli* (Edwards). *a-b,* adult: *a,* male terminalia; *b,* phallosome. *c,* larva: head, mentum and terminal segments. (*a, c,* after Dobrotworsky.)

1, 2-4 branched; 2, 7-12 branched; 3, single; 4, of 14-15 tufts. Anal papillae narrow, about the same length as saddle.

BIOLOGY

C. hilli breeds in subterranean waters in the tunnels of land cray-fish. After overwintering in the larval stage the spring generation usually appears in October. It is one of the common species of *Culiseta*. Mating occurs during the day, males flying about close to the ground in 'searching flight' and attacking approaching females; it occasionally swarms about 2-3 feet above the ground. In the laboratory eggs are laid singly above water level.

It is a day-biting species which attacks man.

DISTRIBUTION

In Victoria *C. hilli* is distributed in the Central Highlands and Gippsland, but it is rare in the Western Highlands. The northern border of distribution lies between the isohyets of 25-30 inches. It also occurs in the south-eastern part of South Australia.

Culiseta (Neotheobaldia) frenchii (Theobald)

Culex frenchii Theobald, 1901, *Mon. Cul.*, 2: 66. *Theobaldia frenchi* Edwards, 1924, *Bull. ent. Res.*, 14: 363. *Culiseta frenchii* Stone, Knight and Starcke, 1959, *Syn. Cat. Mosquitoes World*, 6: 222. Dobrotworsky, 1954, 79: 72.

Similar to *C. hilli* but, in most cases, can be recognized by its larger size and by having dark brown upright scales on the vertex and a pale scaled venter.

ADULT FEMALE

Upright scales on vertex dark brown. Proboscis and palps clothed with dark brown scales. Integument reddish. Scutum clothed with pale golden scales, which are rather smaller than those of *C. hilli*. Posterior pronotum with narrow pale scales. One or two small, pale spiracular bristles. Mesepimeron with patch of scales towards middle; 2 lower bristles. Tarsi (fig. 74 c) with last 2-3 segments pale. Venter pale scaled.

ADULT MALE

Palps slightly longer than proboscis with labella. *Terminalia* (fig. 74 a, b). Coxite almost three times as long as broad. Basal lobe about two-thirds length of coxite, separated only towards tip, which bears tuft of rather short hairs. Paraproct with 4 large and 3 small teeth. Lobes of IXth tergite with 9-15 curved fine setae.

LARVA (fig. 74 d)

Milky-white with black setae. Head and siphon light brown, siphon darker towards end. Antenna slightly shorter than head; seta 1 arising two-thirds of length from base of antenna, 8-12 branched. Head setae: 4, 5-8 branched; 5, 8-11 branched; 6, 7-9 branched; 7, 10-11 branched; 8, single; 9, 3-5 branched. Prothoracic setae: 1, 2, 4, 5 and 6, single; 3 and 7, 2 branched; sometimes seta 4 is 2 branched and seta 7, 3 branched. *Abdomen*. VIIIth segment: lateral comb large patch of fringed

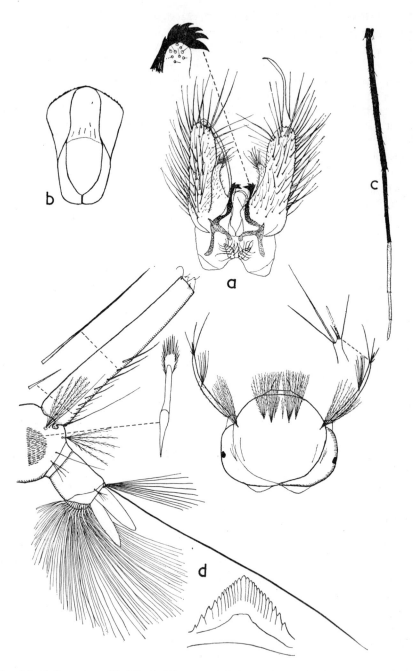

Fig. 74 *Culiseta frenchii* (Theobald). *a-c*, adult: *a*, male terminalia; *b*, phallosome; *c*, hind tarsus. *d*, larva: head, mentum and terminal segments. (*a*, *d*, after Dobrotworsky.)

scales; seta 1, 5-6 branched; 2 and 4, single; 3, 5 branched; 5, 2-3 branched. Siphon cylindrical, tapering towards end; index 5·0-6·2; basal seta single; pecten teeth replaced by 9-12 hairs. Anal segment: saddle, complete ring; seta 1, 4-5 branched; 2, 9-12 branched; 3, single; 4, of 14-15 tufts. Anal papillae narrow, pointed, longer than saddle.

BIOLOGY

C. frenchii is a homodynamous species. It breeds in subterranean water in the tunnels of land cray-fish and both adults and larvae may be found during the winter. It is one of the common species of *Culiseta* and is particularly abundant in upland forests with high rainfall. In the laboratory eggs are laid singly above water level.

It is a day-biting species which attacks man. It is the most troublesome species in forests with an annual rainfall in excess of 50 inches.

DISTRIBUTION

C. frenchii is restricted to the Eastern Highlands of Victoria and Gippsland. The north-western border of distribution is approximately defined by the annual isohyet of 40 inches.

Culiseta (Neotheobaldia) frenchii atritarsalis (Dobrotworsky)

Theobaldia frenchi atritarsalis Dobrotworsky, 1954, *Proc. Linn. Soc. N.S.W.*, 79: 73. *Culiseta frenchii atritarsalis* Stone, Knight and Starcke, 1959, *Syn. Cat. Mosquitoes World*, 6: 222.

ADULT

In general it is darker than the typical *C. frenchii*. Integument brown; scales and proboscis, palps and legs almost black; tarsi black, without pale reflection. Male terminalia are identical with those of the typical *C. frenchii*.

LARVA

The larvae have been bred from eggs laid in the laboratory and are identical with those of the typical form.

DISTRIBUTION

The subspecies is distributed in the South Gippsland highlands (Stony Creek, Boolarra, Kalimna, Hiawatha, Tarra Valley, Drouin West).

Genus CULEX Linnaeus

Culex Linnaeus, 1758, *Syst. Nat. Ed.*, 10: 602.

Of the five subgenera of *Culex* recorded on the Australian mainland only three are represented in Victoria: *Neoculex* with four species, *Lophoceraomyia* with two species and *Culex* with two species and three subspecies.

Characters of the Genus

ADULT

Head with narrow decumbent scales and upright forked scales on vertex. Palps in Victorian species are in the male longer than proboscis, in the female shorter. Antennae of male plumose with hairs of verticils evenly spread round the segments. Spiracular bristles absent. Lower mesepimeral bristles few in number or absent. Pleura usually with small patches of scales, never very densely scaled. Claws of female always simple. Pulvilli (fig. 6 B) always present. Base of subcosta clothed on underside with scales. VIIIth abdominal segment of female short and broad; cerci always short. *Male terminalia.* Coxites without basal lobe or harpago; subapical lobe with several more or less modified bristles; style usually sickle-shaped; paraprocts with tuft of hairs or spines or transverse comb of teeth at tip; phallosome divided into pair of plates with teeth and processes.

LARVA

Antennal tuft usually beyond middle. Thoracic setae usually long and branched; setae 1-3 set on distinct sclerotized plate. Abdomen without dorsal chitinous plates except on anal segment, which is ringed by plate. Siphon long with well developed pecten and several pairs or median row of ventral tufts, and often also with lateral and dorsal tufts. Seta 3 of anal segment long and usually simple.

EGGS

Long, narrowed at one end, with attached corola; always laid as raft.

Division of the genus *Culex* into subgenera is based mainly on the characters of the male terminalia. No satisfactory subgeneric morphological traits have been found in females or larvae.

KEYS TO VICTORIAN SPECIES AND SUBSPECIES OF THE GENUS CULEX

Adult Females

1	Abdomen unbanded	*pseudomelanoconia*
	Abdomen banded	2
2(1)	Tarsi banded	3
	Tarsi unbanded	4
3(2)	Proboscis banded, postspicular area without setae	*annulirostris*
	Proboscis without band, postspiracular area with setae	*postspiraculosus*
4(2)	Tergites with basal bands	5
	Tergites with apical bands	*fergusoni*
5(4)	Wing scales moderately long and dense	6
	Wing scales sparse	10
6(5)	Lower mesepimeral bristles absent	*douglasi*
	Lower mesepimeral bristles present	7
7(6)	Basal tergal bands not constricted laterally	8
	Basal tergal bands constricted and separated from lateral spots at least on tergites 2-5	9
8(7)	Tergites almost black with broad creamy bands. Ventral side of proboscis with pale scales over entire length	*globocoxitus*
	Tergites brown, bands pale. Ventral surface of proboscis with dark scales at tip	*molestus*

9(7) Dark. Postspiracular area with a few scales. Venter with median and lateral patches of black scales .... *australicus*
Lighter. Postspiracular area without scales. Patches on venter usually absent, or if present, small and inconspicuous ..... *fatigans*
10(5) Upright head scales black at sides, lateral flat scales all white
.... *orbostiensis*
Upright head scales pale, some lateral flat scales dark .. *cylindricus*

Larvae (fourth-stage)

1 Skin densely covered with minute spicules .... *fergusoni*
Skin without minute spicules 2
2(1) Head seta 6 single or not more than 2 branched 3
Head seta 6 with more than 3 branches 6
3(2) Head setae 5 a small tuft, 6 single, long 4
Head setae 5 long, single or 2 branched 5
4(3) Surface of saddle covered with short arched rows of fine spines
.... *douglasi*
Surface of saddle covered with strong and irregularly arranged denticles .... *pseudomelanoconia*
5(3) Pecten spines unidentate .... *postspiraculosus*
Pecten spines multidentate .... *orbostiensis* or *cylindricus*
6(2) Six or more pairs of ventral tufts on siphon .... *annulirostris*
Not more than four pairs of ventral tufts 7
7(6) Seta 2 of VIIIth abdominal segment 3 branched; anal papillae third to half length of saddle .... *globocoxitus*
Seta 2 of VIIIth abdominal segment single or 2 branched; anal papillae longer than saddle 8
8(7) Siphon index usually more than 5·0 .... *australicus*
Siphon index less than 4·8 .... *fatigans* or *molestus*

Subgenus NEOCULEX Dyar

Neoculex Dyar, 1905, *Proc. ent. Soc. Wash.*, 7: 45. *Maillotia* Theobald, 1907, *Mon. Cul.*, 4: 274. *Eumelanomyia* Theobald, 1910, *Mon. Cul.*, 5: 114. *Protomelanoconion* Theobald, 1910, *Mon. Cul.*, 5: 462.

Characters of the Subgenus

No lower mesepimeral bristles. *Male terminalia.* Coxite without scales. Paraproct with either a hair tuft at tip or a transverse comb. Phallosome simple, formed of a pair of plates. IXth tergite small.

APICALIS-GROUP

All the known Victorian species of this subgenus belong to the *Apicalis*-group which Edwards (1932) characterized as follows: 'Palpi of male not or scarcely shorter than proboscis and usually more or less hairy. All decumbent scales of vertex narrow. Usually at least some trace of apical pale markings on abdominal tergites'.

At the time only two Australian species, *C. fergusoni* and *C. pseudomelanoconia,* were known, but several new species have since been described and some of these have only basal bands on the tergites. Of the Victorian species, *C. fergusoni* has close relationships with the Holarctic *C. apicalis* Adams; the other species have Oriental affinities.

Culex (Neoculex) fergusoni (Taylor)

Culicada fergusoni Taylor, 1914, *Proc. Linn. Soc. N.S.W.,* 39: 459. *Culex fergusoni* Edwards, 1924, *Bull. ent. Res.,* 14: 398. Dobrotworsky, 1956, 81: 105.

This is the only Victorian species of *Culex* with apical bands on the tergites.

ADULT FEMALE

Upright forked scales on vertex pale. Palps slightly less than quarter length of proboscis. Scutum clothed with small golden scales; line of black scales along acrostichal bristles; supra-alar area with black scales. Posterior pronotum with goldish narrow scales. Mesepimeron with patch of white scales near middle and patch of pale hairs on upper part. No prealar scales or postspiracular scales. Wing membrane with darkened areas. Tarsi brown. Tergites (fig. 75 d) black scaled, with pale apical bands medially constricted or broken on tergites III-V. Venter black scaled basally and pale yellow apically.

ADULT MALE

Palps exceeding length of proboscis with labella by terminal segment and about half of penultimate; apical part of shaft and last two segments with dense long hairs. *Terminalia* (fig. 75 a-c). Coxite not swollen; subapical lobe well divided; distal division with six or seven setae, the first seta slightly flattened, the second, the longest, with hooked tip, the remainder hooked and barbed apically. Proximal division with two long rods with curved flaps apically. Style stout, slightly curved. Paraproct with an apical comb of 11-12 rather long blunt spines. Lateral plates of phallosome without teeth or tubercles, joined by a narrow bridge near tip. Lobes of IXth tergite with 4-5 fine setae.

LARVA (fig. 75 e)

Head broad. Antenna long and curved, pale at middle, dark at base and beyond seta 1; seta 1, 30-35 branched. Head setae: 4, single; 5 and 6, 2 branched; 7, 10 branched; 8 and 9, 3-5 branched. Larval skin densely covered with minute spicules. Prothoracic setae: 1, 2, 3, 5 and 6, single; 3 may be 2 branched; 4, 2 branched; 7, 3-5 branched. *Abdomen.* VIIIth segment: lateral comb of about 60 fringed scales; seta 1, 5-6 branched; 2, 1-2 branched; 3, 7-10 branched; 4, single; 5, 3-4 branched. Siphon long and slender with index about 10; pecten of 14-15 unidentate spines; 10-12 ventral tufts which may be simple or branched. Anal segment: seta 1, 2 branched; 2, 4-5 branched; 3, single; 4, of 12 tufts, 1 or 2 precratal. Anal papillae narrow, pointed; ventral pair about half length of saddle; dorsal pair shorter.

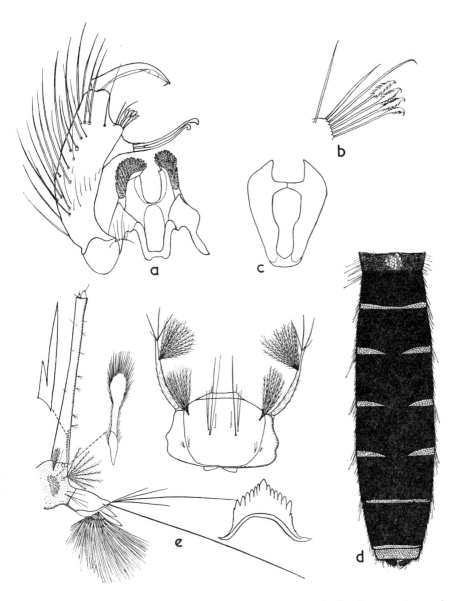

Fig. 75 *Culex fergusoni* (Taylor). *a-d*, adult: *a*, male terminalia; *b*, appendages of subapical lobe; *c*, phallosome; *d*, female abdomen, dorsal view. *e*, larva: head, mentum and terminal segments.

Eggs

The egg raft consists of 84-150 eggs. Anteriorly the egg is almost cylindrical but the posterior third is bent and sharply tapering. The corolla is funnel-like.

BIOLOGY

C. fergusoni is a homodynamous species. Males and females (often gravid) as well as the immature stages can be collected at any time during the winter months at lower altitudes; at higher altitudes the low winter temperatures prevent breeding of this species. It is numerous in the hill forests of Victoria where it breeds in clean water pools and swamps. At lower altitudes and in more exposed situations it is confined to rocky valleys where the larvae are found in back-water pools shaded by trees or overhanging rocks. The larvae are sometimes found in swamps overgrown with bullrushes.

In the laboratory *C. fergusoni* deposited egg rafts on moist filter paper 2-5 inches above the water level. The rafts were not stuck to the paper for they floated away when the water level was raised. When the water level was kept low, the rafts, if on a vertical surface, dropped into the water as the larvae pushed off the egg caps. When the rafts were placed on an inclined surface, the newly hatched larvae crawled over the moist filter paper to the water.

C. fergusoni is not a man-biting mosquito.

DISTRIBUTION

In Victoria the distribution of *C. fergusoni* is almost identical with that of *Culiseta inconspicua* and both species usually share the same breeding places. It is distributed in Eastern Australia from southern Queensland to South Australia and Tasmania.

Culex (Neoculex) douglasi Dobrotworsky

Culex douglasi Dobrotworsky, 1956, *Proc. Linn. Soc. N.S.W.*, 81: 105-114.

It is readily distinguished from all other Victorian species by the presence of basal white bands on both tergites and sternites.

ADULT FEMALE

Upright forked scales on vertex, pale. Proboscis and palps blackish with violet reflections; palpi about one-sixth length of proboscis. Scutum clothed with golden scales. Posterior pronotum with a few pale narrow curved scales. Sternopleuron with two small patches of broad small scales. Mesepimeron with patch of pale hairs and a few scales on upper part and usually a small patch of scales near middle; no prealar scales. Wing scales brown; plume scales very narrow. Legs brown scaled; tarsal claws of all legs small and simple. Abdomen black scaled with white basal bands on tergites and sternites; black scales on venter with violet-pink reflections.

ADULT MALE

Palps (fig. 76 c) longer than proboscis with labella by terminal segment and about a third or half of penultimate; last two segments clothed with long hairs, shaft with about 6 hairs at apex. *Terminalia* (fig. 76 a, b). Coxite narrow, subapical lobe well divided; distal division with two long and three shorter bristles with recurved tips, three shorter ones barbed apically. This division of lobe accompanied by long seta. Proximal division with two long sinuous rods curved, and flattened apically. Style stout, slightly curved, broadening slightly distally and narrowing abruptly before tip; appendage short. Paraproct with apical comb of 10-12 rather long, blunt spines. Lateral plates of phallosome joined by narrow bridge near tip. Lobes of IXth tergite with 4-5 rather long setae.

LARVA (fig. 76 d)

Larva pale except metathorax and abdominal segments 1, 2, 3 and 5, which are dark dorsally. Head broad. Antennae long, curved, white and strongly spiculated up to seta 1, dark above it; seta 1, about 30 branched. Head setae: 4 and 6, single; 5, small tuft, 4-5 branched; 7, 9-10 branched; 8 and 9, 2-4 branched. Mentum with central tooth and 5 lateral teeth, rarely 6 or 7. Prothoracic setae: 1-6, single; 7, 2 branched. *Abdomen*, VIIIth segment: lateral comb of about 40 fringed scales; seta 1, 3-5 branched; 2, 2 branched; 3, 7-8 branched; 4, single; 5, 3-4 branched. Siphon slender with dark ring towards the middle; index 6·0-7·1; pecten

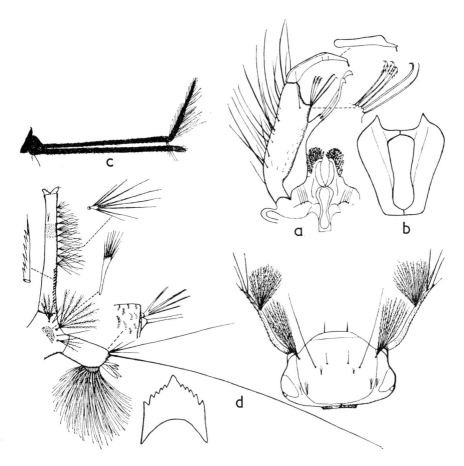

FIG. 76 *Culex douglasi* Dobrotworsky. *a-c*, adult: *a*, male terminalia; *b*, phallosome; *c*, proboscis and palp of male. *d*, larva: head, mentum and terminal segments.

of 16-21 spines, each with 5-7 denticles; 8-12 ventral tufts each of 6-8 branches; dorsolateral seta 2-3 branched. Anal segment: surface of saddle covered with short arched rows of fine spines; seta 1, small, 3-4 branched; 2, with 1 long and 2-3 short branches; 3, single; 4, of 14 tufts, one precratal. Anal papillae narrow, pointed; dorsal pair shorter than ventral and about one-third length of saddle.

EGGS

The egg rafts are oval or rounded; after laboratory feeding on human blood they contained up to 48 eggs. The eggs are thick anteriorly, bent and sharply tapering posteriorly. The funnel-like corolla is filled with a sticky substance.

BIOLOGY

The adults usually rest in dark recesses in the banks of streams or in rocks; those collected late in April or in May had a well developed fat body. They could be induced to feed upon human blood after exposure to artificial lighting for several days at a temperature of 77°F. Oviposition commenced after the fourth blood-meal but each subsequent meal was followed by egg laying. In the laboratory egg rafts were deposited on moist filter paper from one to three inches above the water surface. When development of the eggs is complete, the larvae hatch immediately the water level is raised up to the raft. Hatching can occur if the water level remains low. In this case the lifting of the egg cap causes the eggs to become detached from the corolla and the raft then drops into the water.

In Victoria the larvae breed in rocky valleys, in shallow, cool, clean backwater pools with a sandy bottom and shaded by rocks or trees. The larvae are most abundant under overhanging stones. The larvae and pupae usually disappear from breeding places late in April. This mosquito rarely attacks man.

DISTRIBUTION

C. douglasi has a patchy distribution in Victoria, but in the vicinity of breeding places it is always numerous; Victorian records are: Weeragua, Delatite, Meredith, Violet Town and Euroa. It has also been recorded from Queensland and New South Wales.

Culex (Neoculex) postspiraculosus Lee

Culex postspiraculosus Lee, 1944, *Proc. Linn. Soc. N.S.W.*, 69: 221.

Easily recognized by its having an unbanded proboscis, banded tarsi and black sternites with white lateral patches.

ADULT FEMALE

Vertex with narrow curved pale scales and black upright forked scales. Proboscis black scaled above, pale below except on apical third which is black. Palps black. Integument black. Scutum clothed with dark bronze scales becoming pale around prescutellar area. Postspiracular area (fig. 77 d) with 4-6 black bristles and a few broad pale scales. Patches of pale scales on sternopleuron among lower bristles and toward middle; on mesepimeron, among upper bristles. Wings and legs dark scaled. Tarsi banded (fig. 77 f). Knob of halteres black with pale scales. Tergites and sternites black scaled; tergites with white basal bands, sternites with lateral patches of white scales.

ADULT MALE

Palps longer than proboscis with labella by terminal segment and almost

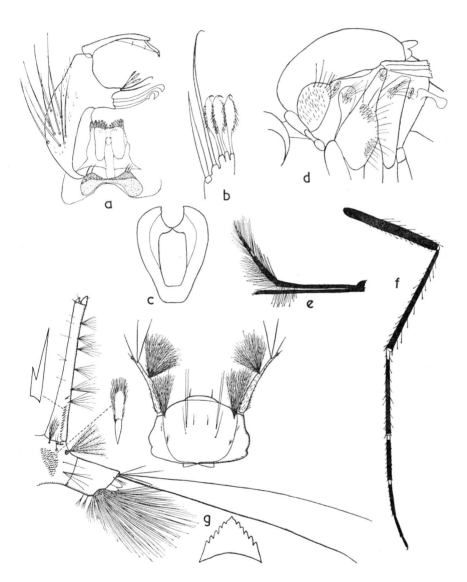

FIG. 77 *Culex postspiraculosus* Lee. *a-f*, adult: *a*, male terminalia; *b*, appendages of subapical lobe; *c*, phallosome; *d*, side view of thorax; *e*, proboscis and palp of male. *f*, hind leg. *g*, larva: head, mentum and terminal segments.

half of penultimate; dark scaled with some pale scales at base and on tip of terminal segment. *Terminalia* (fig. 77 a-c). Coxite triangular in shape. Subapical lobe well divided; distal division with 5 setae, this division of lobe accompanied by a long seta. Proximal division with two long sinuous rods curved and flattened apically. Style stout, narrowing abruptly before tip. Paraproct with apical comb of 8-9 rather long, blunt spines. Lateral plates of phallosome joined by narrow bridge near tip. Lobes of IXth tergite with 1-3 setae.

LARVA (fig. 77 g)

Dark, with white area in mid-part of antenna. Head broad. Antenna long; seta 1, 23-24 branched. Head setae: 4, 6 and 9, 2 branched; 5, single; 7, 15-17 branched; 8, 3-4 branched. *Abdomen.* VIIIth segment: lateral comb of about 80 fringed scales; seta 1, 4-6 branched; 2 and 4, single; 3, 8 branched; 5, 4-5 branched. Siphon long with index about 6-7; 6 pairs of lateral tufts and 3 dorsolateral pairs; pecten of about 16 spines. Anal segment: seta 1, 2 branched; 2, with one long and 3 short branches; 3, single; 4, of 13-14 tufts, 3 precratal.

BIOLOGY AND DISTRIBUTION

Larvae have been found in open swamps overgrown with reeds. In Victoria it has been found only in the Bairnsdale area. It is also recorded from southern Queensland and New South Wales.

Culex (Neoculex) pseudomelanoconia Theobald

Culex pseudomelanoconia Theobald, 1907, *Mon. Cul.*, 4. 416.

This is a dark scaled species without pale bands on legs or abdomen.

ADULT FEMALE

Vertex with narrow curved and upright forked creamy scales. Proboscis and palps black scaled. Scutal integument brown. Scutum with narrow curved dark-golden scales. Posterior pronotum with a few narrow curved golden scales. Patch of scales near middle of sternopleuron and some scales along its posterior border; a few scales at base of upper mesepimeral bristles. No lower mesepimeral bristles. Wings and legs dark scaled. No distinct knee spots. Tergites black scaled. Sternites brown scaled.

ADULT MALE

Palps exceed proboscis with labella by terminal segment and about half penultimate; shaft apically with a few moderately long hairs; penultimate and terminal segments with long but not dense hairs. *Terminalia* (fig. 78 a-c). Coxite not swollen; subapical lobe well divided; distal division with five setae; the first seta straight, the second, the longest, with hooked tip, the remainder hooked and barbed apically. Proximal division with two long rods with curved flaps apically. Style stout. Paraproct with apical comb of 12-13 rather long blunt spines. Lateral plates of phallosome without teeth or tubercles, joined by a narrow bridge near tip. Lobes of IXth tergite with 3-5 setae.

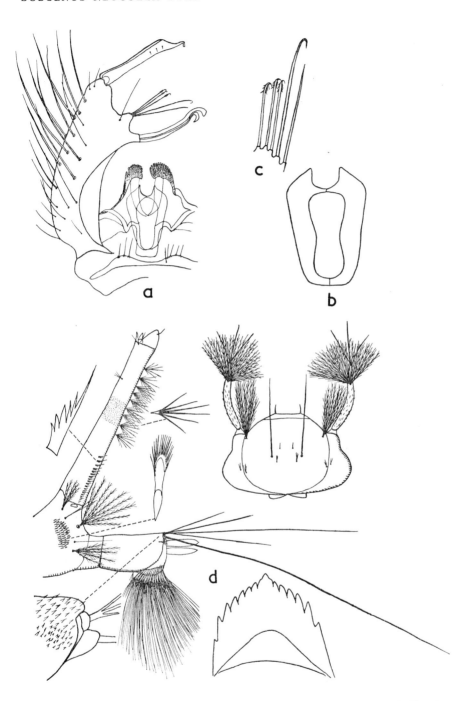

FIG. 78 *Culex pseudomelanoconia* Theobald. *a-c*, adult: *a*, male terminalia; *b*, phallosome; *c*, appendages of subapical lobe. *d*, larva: head, mentum and terminal segments.

LARVA (fig. 78 d)

Larva dark except anterior two-thirds of thorax and IVth abdominal segment, which are pale. Head broad. Antenna long curved, strongly spiculate; seta 1, about 30 branched. Head setae: 4 and 5 very small; 4, single or 2 branched; 5 and 7, 2-5 branched; 6, single, long. Mentum with large central tooth and 6-7 lateral teeth. *Abdomen*. VIIIth segment: lateral comb patch of about 45 fringed scales; seta 1, 4-5 branched; 2, 2 branched; 3, 8-9 branched; 4, single; 5, 4-5 branched. Siphon slender with dark ring near middle, index about 6; pecten of 13-19 spines, each with 5-7 denticles; 9-10 ventral tufts each of 6-9 branches; dorsolateral seta 2-3 branched. Anal segment: surface of saddle covered with irregularly arranged denticles on dorsodistal part; seta 1, small, 3-4 branched; 2, 4 branched; 3, single; 4 (ventral brush) of 14-15 tufts, one precratal. Anal papillae narrow, pointed, equal, about half of length of saddle.

BIOLOGY AND DISTRIBUTION

Larvae have been found in ground pools in eucalypt forests which are shaded for part of the day. The water is clear. In Victoria it has been found only at Weeragua (East Gippsland). It is also recorded from New South Wales and Queensland.

Subgenus LOPHOCERAOMYIA Theobald

Lophoceraomyia Theobald, 1905, *J. Bombay nat. Hist. Soc.*, 16: 245 (April 15). *Lophoceratomyia* Theobald, 1905, *Ann. hist. nat. Mus. Hung.*, 3: 93 (June, *lapsus*). *Philodendromyia* Theobald, 1907, *Mon. Cul.*, 4: 623. *Cyathomyia* Meijere, 1910, *Ann. bot. Gdn. Buitenz.*, Suppl. 3: 921.

Characters of the Subgenus

Antennae (fig. 79 a) of male with thin matted hair-tuft on segment 9, usually also with tufts of modified scales on segments 6-8. Vertex usually with small flat scales in front, towards eye margins. Lower mesepimeral bristles present. VIIIth tergite of male deeply emarginate. Wing scales usually very scanty, except towards tip. Terminalia similar to that of subgenus *Neoculex*.

GROUP A (MINUTISSIMUS-GROUP)

Characters of the Group

Basal segments of male antenna simple; segments 6-8 without scale tufts. Palps of male simple.

Culex (Lophoceraomyia) cylindricus Theobald

Culex cylindricus Theobald, 1903, *Mon. Cul.*, 3: 202. Dobrotworsky, 1957, 82: 319.

The upright scales on the vertex all pale; some lateral flat scales dark, the tergites with white basal bands; the sternites pale scaled; the tarsi unbanded.

Adult Female

Upright forked scales on vertex all pale. Proboscis black scaled. Scutum clothed with silvery scales with light-goldish gleam, only slightly curved. Posterior pronotum bare. Sternopleuron with patch of scales towards posterior edge and in middle. Mesepimeron with 1 lower bristle; a few scales below upper bristles. Tergites black scaled with white basal bands; sternites pale scaled.

Adult Male

Palps exceeding length of proboscis with labella by about length of terminal segment; shaft with 4-5 long hairs at apex; last two segments

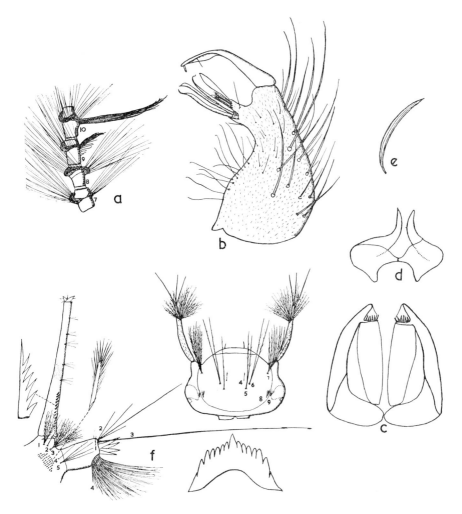

Fig. 79 *Culex cylindricus* Theobald. *a-e*, adult: *a*, segments 7-10 of male antenna; *b*, coxite of male terminalia; *c*, paraprocts; *d*, phallosome; *e*, scale from scutum. *f*, larva: head, mentum and terminal segments. (after Dobrotworsky.)

sparsely clothed with long hairs. Antennal segment 9 (fig. 79 a) with short tuft of matted hairs; 10 with long tuft of thick matted hairs. *Terminalia* (fig. 79 b-d). Coxite with about a dozen long bristles on inner side; subapical lobe of coxite with 1 leaf, 5 more or less flattened setae and 3 long rods; style stout, curved. Paraproct with small number of short spines. Lateral plates of phallosome pointed and bent, joined by narrow bridge near base. Lobes of IXth tergite with 4-8 setae.

LARVA (fig. 79 f)

Similar to that of *C. orbostiensis* and no morphological traits have been found for separating the larvae of these species. In Victoria, however, the presence of a conspicuous dark band across the middle of the siphon of larva of *C. cylindricus* and its absence from the larvae of *C. orbostiensis* permits the recognition of the two species.

BIOLOGY

In the Mildura area it has been recorded biting cattle.

DISTRIBUTION

In Victoria *C. cylindricus* has been recorded only from the Mildura area. It also occurs in New South Wales and Queensland.

GROUP B (FRAUDATRIX-GROUP)

Characters of the Group

Basal segments of male antenna simple; segments 6-8 with more or less conspicuous scale tufts. Palps of male often with a pair of finger-like projections at base.

Culex (Lophoceraomyia) orbostiensis Dobrotworsky

Culex orbostiensis Dobrotworsky, 1957, *Proc. Linn. Soc. N.S.W.*, 82: 317.

The females are very similar indeed to *C. cylindricus* and can be distinguished by the slightly narrower scales on the scutum and the colour of the head scales.

ADULT FEMALE

Upright scales on vertex black at sides, lateral flat scales all white. Proboscis clothed with dark brown scales becoming black towards base and apex. Palps about one-eighth length of proboscis. Scutum clothed with small curved dark-golden scales (fig. 80 e) becoming paler around bare area. Posterior pronotum bare. Sternopleuron with patch of scales towards posterior edge and in middle. Mesepimeron with single lower bristle and patch of scales on upper part. Tarsal claws of all legs simple. Tergites with white basal bands. Venter pale scaled.

ADULT MALE

Proboscis with about 15 long upright hairs dorsally towards apex. Palps dark scaled, exceeding length of proboscis with labella by terminal segment; first segment with two hairy apical processes; shaft with a few long hairs at apex; last two segments with long, dark hairs. Antennal segments 6 and 10 (fig. 80 a) with tuft of long, narrow scales; 7 and 8 with short wavy tuft of modified hairs; 9 with tuft of long, subapically curved hairs. *Terminalia* (fig. 80 b-d). Coxite with inner row of 4 long bristles; subapical lobe with 3 long rods, 2 leaves and 5 more or less flattened bristles; style stout, curved, broadening distally. Paraproct with small number of spines. Lateral plates of phallosome sharply pointed and bent, without teeth, and joined by narrow bridge near base. Lobes of IXth tergite with 5-6 setae.

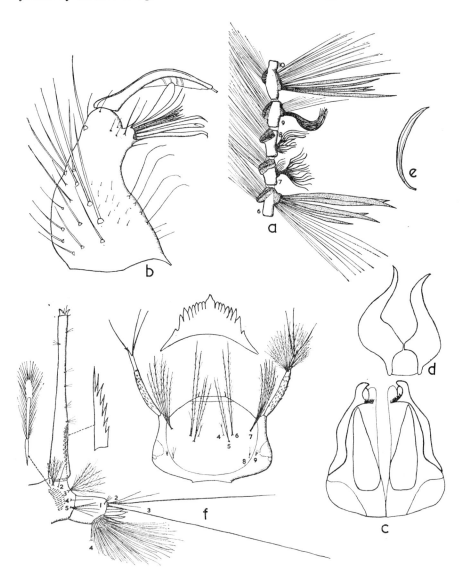

Fig. 80 *Culex orbostiensis* Theobald. *a-e*, adult: *a*, segments 6-10 of male antenna; *b*, coxite of male terminalia; *c*, paraprocts; *d*, phallosome; *e*, scale from scutum. *f*, larva: head, mentum and terminal segments. (after Dobrotworsky.)

LARVA (fig. 80 f)

Head broad. Antenna long, strongly spiculate; seta 1 with about 30 plumose branches; subterminal setae strong, slightly removed from tip. Head setae: 4, fine, short, with 2-4 branches; 5, long, 2-3 branched; 6, longer than 5, 2 branched; 7, long, 6-7 branched; 8, 3-4 branched; 9, 3 branched. Mentum with broad central tooth and 8 lateral teeth on each side. Prothoracic setae: 1 and 2, single, long; 3, 2 branched, about one-quarter length of 1 and 2; 4, 2 branched, twice as long as 3; 5 and 6, single, long; 7, 3 branched, about two-thirds as long as 6; 8, 2 branched, about the same length as 7. *Abdomen*. VIIIth segment: lateral comb of about 50 fringed scales; seta 1, 4-5 branched, slightly plumose; 2, single; 3, 6-7 branched, plumose; 4, single; 5, 5 branched, slightly plumose. Siphon slender, index $8 \cdot 0$-$9 \cdot 0$. In some Queensland specimens the siphon has a faint dark band just beyond middle. Pecten of 14-16 spines, each with several denticles; 5 pairs of ventral tufts each of 2-4 branches; tufts usually shorter than diameter of siphon at their points of attachment. Anal segment: saddle complete; setae: 1, 3-4 branched; 2, with one long and 3 short branches; 3, single, long; 4, of 11-12 tufts, each 3-7 branched. Anal papillae almost equal in size, slightly less than half length of saddle.

BIOLOGY

It has several generations during the summer; the first generation of adults in the Melbourne area appears early in September. It is a day-biting mosquito.

DISTRIBUTION

In Victoria *C. orbostiensis* is common only in East Gippsland but it has also been recorded from Woori-Yallock, Maroondah and Christmas Hills. It also has been recorded from Queensland.

Subgenus CULEX Linnaeus

Culex Linnaeus, 1758, *Syst. Nat. Ed.*, 10: 602. For synonyms see Stone, Knight and Starke (1959).

Characters of the Subgenus

Antennae of male unmodified. No broad scales on vertex around eye margins. Lower mesepimeral bristles present or absent. VIIIth tergite of male not very deeply emarginate. Terminalia: coxite without scales; style without subapical spiny crest; paraproct with dense tuft of hairs or spines at tip; phallosome usually complicated, with various processes and teeth. Wing scales nearly always dense.

Edwards (1932) divided species of the subgenus *Culex* into two groups.

Sitiens group. Most species of this group are distributed in the Oriental Region but some occur in Australia and Africa; one species, *C. annulirostris*, is found in Victoria.

Pipiens group. Most species of this group are distributed in the Ethiopian Region. All Nearctic and Neotropical members of the subgenus belong to this group and a few species occur in other regions. In Victoria this group is represented by *C. globocoxitus* and three subspecies of *C. pipiens*: *C.p. fatigans, C.p molestus* and *C.p. australicus*.

GROUP A (SITIENS-GROUP)

Characters of the Group

Lower mesepimeral bristles absent. Proboscis with white band in middle. Tarsi banded.

Culex (Culex) annulirostris Skuse

Culex annulirostris Skuse, 1889, *Proc. Linn. Soc. N.S.W.*, 3: 1737. *Culex jepsoni* Bahr (*non* Theobald), 1912, *J. Lond. Sch. Trop. Med.*, 1: 18. *Culex palpalis* Taylor, 1912, *Bull. Northern Terr.*, la: 12. *Culex somerseti* Taylor, 1912, *Rept. Commis. publ. Health* Qd, p. 28. *Culicelsa consimilis* Taylor, 1913, *Austr. Inst. Trop. Med. Rept.*, p. 8. *Culicelsa simplex* Taylor, 1914, *Trans. ent. Soc.*, p. 698.

The proboscis and tarsi are banded. The tergites have white bands expanding mesially; the sternites are white scaled with lateral apical black patches.

ADULT FEMALE

Vertex clothed with narrow curved pale scales; upright scales dark in front and laterally, white medially. Proboscis (fig. 81 c) black with white band near middle; palps dark with white tip. Scutum clothed with bronze scales becoming pale in the scutal angle and round the prescutellar area. Posterior pronotum with bronze scales becoming pale below. Sternopleuron with two patches of white scales, mesepimeron with one. Wing dark scaled. Legs (fig. 81 d) dark; femora mottled; some pale scales along tibiae; tarsi with white bands. Tergites (fig. 81 e) black with white basal bands widened in middle on II-IV. Sternites (fig. 81 f) white scaled with apical lateral patches of black scales.

ADULT MALE

Palps longer than proboscis with labella by terminal segment; dark scaled, with white bands at middle of segment 3 and at base of segments 4 and 5; 5th segment white apically; shaft with numerous long hairs on apical half. Scutal scales lighter than in female. *Terminalia* (fig. 81 a, b). Subapical lobe of coxite with 3 long rods with curved tips; 3 slightly flattened setae and a leaf. Style short and stout, sickle-shaped. Paraproct with crown of spines, without basal arm. Dorsal processes of mesosome strong, pointed, directed outwards; ventral processes with spiculated tip and large process with 4 blunt teeth. Lobes of IXth tergite with 7-9 setae.

LARVA (fig. 81 g)

Head broad. Antennae pale, dark at base and above seta 1, which is large with 35-40 branches. Head setae: 4, single; 5, 3-6 branched; 6, 2-4 branched; 7, 10-12 branched; 8, 2-3 branched; 9, 3-5 branched. Prothoracic setae: 1, 2, 3, 5 and 6, single; 4, 2 branched; 7, 3 branched. *Abdomen.* VIIIth segment: lateral comb of 30-43 fringed scales; seta 1, 5 branched; 2 and 4, single; 3, 7-9 branched; 5, 4-6 branched. Siphon long and slender with index 7·0-8·4; pecten of 8-13 spines; 6 pairs of ventral tufts. Anal segment: seta 1, 2-3 branched; 2, 3-4 branched; 3,

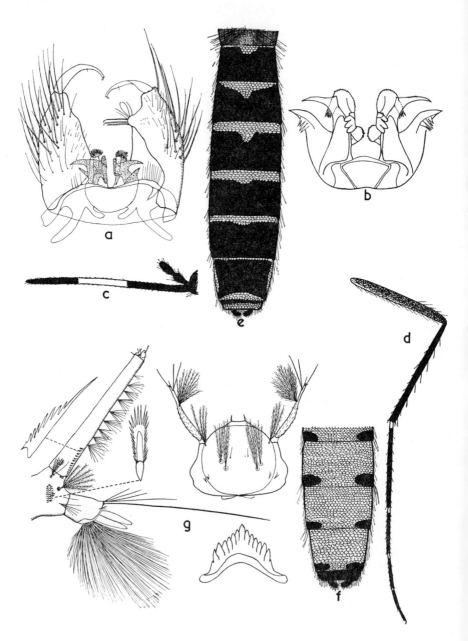

Fig. 81 *Culex annulirostris* Skuse. *a-f*, adult: *a*, male terminalia; *b*, phallosome; *c*, female proboscis and palp; *d*, hind leg; *e*, female abdomen, dorsal view; *f*, terminal segments of female abdomen, ventral view. *g*, larva: head, mentum and terminal segments.

single; 4, of 11-12 tufts. Anal papillae narrow, pointed, about two-thirds length of saddle.

BIOLOGY

C. annulirostris breeds in fresh and brackish ground pools, swamps, ditches and channels. It is a very vicious mosquito, which may bite during the day but is particularly active after sunset. It attacks man and a wide variety of animals and birds.

DISTRIBUTION

In Victoria *C. annulirostris* is abundant north of the Central Highlands; south of them it is not common and appears mainly during late summer and in the autumn. It also occurs through Australia and in Melanesia, Polynesia and Papua.

GROUP B (PIPIENS-GROUP)

All but one of the representatives of this group in Australia (Victoria) are members of the *Culex pipiens* complex, which has a world-wide distribution and presents a difficult taxonomic problem (Mattingly et al. 1951; Laven, 1951, 1953, 1959; Dobrotworsky and Drummond, 1953; Kitzmiller, 1953, 1959; Dobrotworsky, 1954; Rozeboom, 1954, 1958; Omori et al., 1955; Bekker, 1956; Barr, 1960).

The most widely distributed members of this complex are *pipiens, fatigans* and *molestus*. In attempting to determine their taxonomic status the main criterion used has been the degree of reproductive isolation as indicated by cross-breeding experiments but the results of such experiments are often difficult to interpret and difficult to harmonize with field observations.

In the United States, *pipiens, fatigans* and *molestus* are able to maintain themselves as distinct populations yet there seem to be no genetic barriers between them (Rozeboom and Gifford, 1954). Differences in insemination rates in mixed laboratory populations were due not to any demonstrable barrier to mating but apparently to different degrees of sexual activity. Rozeboom and Gifford suggested that differences in sexual activity may serve to maintain the identity of populations even in regions where the ranges of the three overlap.

The status of *molestus,* in particular, is obscure; over wide areas in various parts of the world it is sympatric with either *fatigans* or *pipiens* and yet, for the most part, retains its identity. Mattingly (1951) regards it as an urban biotype of *pipiens,* a view that is supported by Kitzmiller (1959) and Barr (1960). Kitzmiller (1959) describes it as representing populations of *pipiens* with high gene frequencies for autogeny and stenogamy and suggests that such concentrations are the result of selection for adaptation to certain domestic conditions.

The complexities of the taxonomic problem are well illustrated by the work of Laven (1951, 1953, 1959) on European and North African strains of *pipiens*. He found five different 'crossing types' of *pipiens* in Europe and North Africa and demonstrated that north German strains, western European, Italian and north African strains behaved as separate species: they showed incompatibility in both directions so that gene flow between them would be absent. On the other hand, a cross between males of a south German strain and females of north and west German strains gave fully fertile offspring but the reverse crosses were sterile. Laven referred to these strains as semi-species and concluded 'that the *C. pipiens* complex consists of many crossing types. These can be full species, semi-species or strains still capable of interbreeding but with cryptic differentiation . . . '

Kitzmiller and Laven (1959) in a discussion of evolutionary mechanisms in mosquitoes have commented on the slight morphological differentiation between species with marked physiological differences and suggest that speciation in mosquitoes may be a result of a low mutation rate for morphological differences but a higher rate for physiological adaptive characters.

In Australia, the complex is represented by *fatigans, molestus* and *australicus*, which will be treated as subspecies of *C. pipiens,* and by *C. globocoxitus*.

C.p. fatigans is widely distributed in Australia but is absent from Gippsland and is present in the Melbourne area only during late summer and autumn. *C.p. molestus* occurs in southern Australia and is most common in the south-eastern part. Apparently it first established itself in the Melbourne area and then spread eastwards and also northwards into territory already occupied by *C.p. fatigans*. In the laboratory the two subspecies in Victoria interbred freely (Dobrotworsky and Drummond, 1953) and natural hybrid populations have been recorded in Melbourne and in several places in northern Victoria. But as in the United States, the two subspecies, even in areas where their ranges overlap, retain their identity. In some cases in Victoria gene flow is prevented when *C.p. molestus* breeds in septic tanks and underground drains; stenogamy and autogeny maintain a pure population. Exploitation of such habitats probably accounts for the cryptic differentation of strains that has been observed in *molestus* both in Europe (Laven, 1953) and in Victoria (Dobrotworsky, 1955). When strains from Moe, Yarram and Point Lonsdale were crossed with a Melbourne strain, offspring were produced only if the males were of the Melbourne strain; reciprocal crosses were sterile due to inviability of larvae.

In Victoria, the spread of *C.p. molestus* has apparently followed the extension of the use of septic tank systems and in this respect it could be regarded as an urban biotype. It is, however, not restricted to cities and towns; it breeds in septic tanks on isolated farms in country areas.

C.p. australicus, unlike *fatigans* and *molestus,* is primarily a rural mosquito and is neither autogenus nor stenogamous. In the laboratory, crosses between female *australicus* and male *fatigans* or *molestus* were invariably sterile. Reciprocal crosses were largely fertile but there was a heavy mortality of the F_1 pupae. The available evidence indicates that, in nature, *C.p. australicus* is reproductively isolated from the other two subspecies.

The remaining member of the *C. pipiens* group in Victoria is *C. globocoxitus*. Due to mating preferences it is almost completely isolated from both *fatigans* and *australicus* but not from *molestus*. Males of *molestus* ignore *globocoxitus* females, but the reverse is not true; in mixed laboratory populations *globocoxitus* males do not distinguish between the females of the two species and crosses are highly fertile. Backcrosses with *molestus* males were fertile but those with *globocoxitus* females were sterile. This potentiality for inter-breeding is occasionally realized when the two species occupy the same breeding site (Dobrotworsky, 1953) but, as a rule, the populations are kept distinct by ecological isolation.

Characters of the Group

Lower mesepimeral bristles present. Proboscis and tarsi usually unbanded.

Culex (Culex) globocoxitus Dobrotworsky

Culex globocoxitus Dobrotworsky, 1953, *Proc. Linn. Soc. N.S.W.*, 77: 357.

The proboscis is pale below to the tip; the tergites are black scaled with creamy bands not constricted laterally.

ADULT FEMALE

Vertex clothed with pale narrow curved scales and upright scales pale medially becoming dark laterally. Palps dark with some pale scales dorsally. Proboscis dark above, pale below. Scutum clothed with golden scales. Posterior pronotum with narrow golden scales. Sternopleuron and mesepimeron each with two patches of pale scales. One lower mesepimeral bristle; no prealar scales on tip of sternopleuron. Wing and legs dark scaled. Tergites (fig. 82 d) black scaled with creamy, unconstricted basal bands. Sternites creamy, sometimes with patches of black scales medially and laterally.

ADULT MALE

Palps (fig. 82 c) longer than proboscis by two-thirds of terminal segment; dark scaled except on underside of segments 1-3, which are pale scaled; hairs on penultimate and terminal segments relatively short and sparse; shaft with about 10 long hairs apically. *Terminalia* (fig. 82 a, b). Coxite very broad and swollen, with dense yellowish setae and bunch of long setae on inner face. Subapical lobe with 3 rods, one of which is short and spine-like; leaf narrow, accompanied by two long setae and one short. Style strong, sickle-shaped, very narrow distally. Paraproct without basal arm. Ventral processes of mesosome narrow and bent outwardly; dorsal processes stout and pointed, directed towards tips of ventral processes. Lobes of IXth tergite with 3-6 setae.

LARVA (fig. 82 e)

Head and siphon light brown. Antenna about five-eighths of length of head; seta 1, with about 23 branches. Head setae: 4, single; 5, 5-7 branched; 6, 3-6 branched; 7, 8-12 branched; 8, 3-5 branched; 9, 4-7 branched. Mentum with larger central tooth and 7-8 lateral teeth. Prothoracic setae: 1, 2, 3, 5 and 6, single; 4, 2 branched; 7, 2-3 branched. *Abdomen*. VIIIth segment: lateral comb of 40-50 fringed scales; seta 1, 4-7 branched; 2 and 4, single; 3, 6-9 branched; 5, 4-6 branched. Siphon

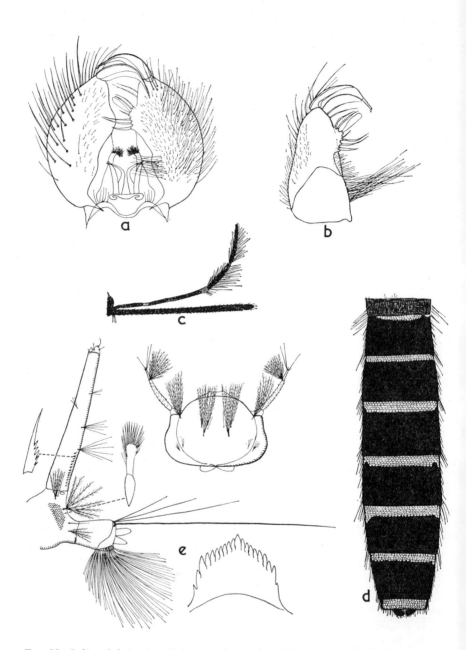

Fig. 82 *Culex globocoxitus* Dobrotworsky. *a-d*, adult: *a*, male terminalia; *b*, coxite from inner lateral aspect; *c*, male proboscis and palp; *d*, female abdomen, dorsal view. *e*, larva: head, mentum and terminal segments.

slender with index 4·6-6·4; pecten of 11-16 spines, 3 pairs of ventral tufts and one pair of dorsolateral tufts. Anal segment: setae 1 and 3, single; 2, usually 3 branched, may be 4-5 branched; 4, about 12 tufts. Anal papillae short, usually half to one-third length of saddle.

BIOLOGY

It breeds in open swamps, and in large and small grassy pools in creek beds; it can tolerate brackish and polluted water. This is a homodynamous and stenogamous species; it mates in the laboratory in a space as small as 3 cubic inches.

The rafts obtained from females fed in nature consist of 190-302 eggs. It is not a man-biting mosquito, but after a human blood meal in the laboratory it deposits rafts containing only 70-100 eggs.

DISTRIBUTION

In Victoria *C. globocoxitus* is a widely distributed species, but is not so numerous as *C.p. australicus*. It is also recorded from south-western Queensland (Dr E. N. Marks, personal communication), New South Wales, South Australia, Western Australia and Tasmania.

Culex (Culex) pipiens australicus Dobrotworsky and Drummond

Culex p. australicus Dobrotworsky and Drummond, 1953, *Proc. Linn. Soc. N.S.W.,* 78: 143.

Usually recognized by the following combination of characters: the tergal bands are constricted laterally; the sternites have conspicuous median, and apical lateral, patches of black scales; the presence of a few white broad scales on the postspiracular area.

ADULT FEMALE

Vertex clothed with narrow pale scales, and upright scales pale medially, black laterally. Palps dark scaled, with a few pale scales apically. Proboscis (fig. 83 c) dark scaled above, pale below except apical quarter which is black. Integument dark brown. Scutum clothed with bronze and golden scales becoming pale round prescutellar area. Posterior pronotum with narrow dark-golden scales and a few white scales below. Postspiracular area with a few white scales. Sternopleuron and mesepimeron each with two patches of white scales; single lower mesepimeral bristle. Wings and legs dark scaled. Tergites (fig. 83 e) black scaled with white basal bands constricted laterally; bands may be separated from white lateral patches on segments II-VI or only on segments II-IV. Sternites (fig. 83 f) white scaled with median and apical lateral patches of black scales.

ADULT MALE

Palps (fig. 83 d) longer than proboscis with labella by terminal segment, dark scaled with some white scales laterally on shaft near middle; streak of white scales on penultimate segment and white patch at base of terminal segment below; shaft with long hairs on apical half. *Terminalia* (fig. 83 a, b). Subapical lobe of coxite with 3 almost equal rods; leaf accompanied by 3 setae, one slightly flattened. Style strong, sickle-shaped, very

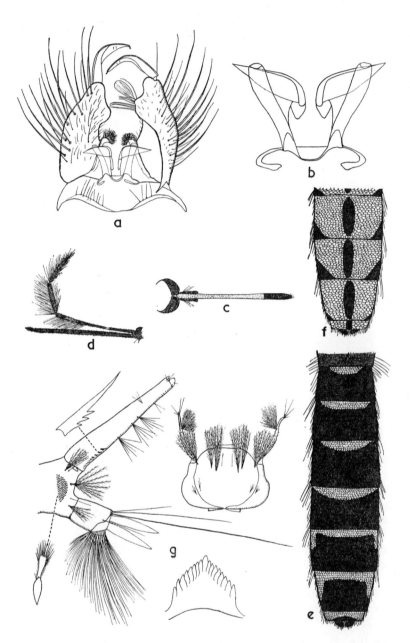

FIG. 83 *Culex pipiens australicus* Dobrotworsky and Drummond. *a-f*, adult: *a*, male terminalia; *b*, phallosome; *c*, proboscis and palp of female, ventral view; *d*, proboscis and palps of male; *e*, female abdomen, dorsal view; *f*, terminal segments of female abdomen, ventral view. *g*, larva: head, mentum and terminal segments. (*b, d*, after Dobrotworsky.)

narrow distally. Paraproct without basal arm; dorsal processes of mesosome transparent, directed outwards, thickened distally and slightly excavated at tip; ventral processes long and leaf-like distally. Lobes of IXth tergite with 5-15 setae.

LARVA (fig. 83 g)

Head and siphon light brown. Antenna long; seta 1, 21-26 branched. Head setae: 4, single; 5, 4-5 branched; 6, 3-5 branched; 7, 6-10 branched; 8 and 9, 3-5 branched. Mentum with large central tooth and 8-9 lateral teeth. Prothoracic setae: 1, 2, 3, 5 and 6, single; 4 and 7, 2 branched. *Abdomen.* VIIIth segment: lateral comb patch of 31-40 fringed scales; seta 1, 5-8 branched; 2 and 4, single; 3, 7-9 branched; 5, 4-5 branched. Siphon with index 4·4-6·4, mean 5·5; pecten of 9-13 teeth; usually 4 pairs of ventral tufts and one dorsolateral pair. Anal segment: seta 1 and 3, single; 2, 2 branched; 4 of 11-12 tufts; anal papillae usually about as long as saddle; larvae collected during spring in some rain-water pools have anal papillae almost twice as long as saddle.

EGGS

The egg rafts are very large, containing up to 503 eggs.

BIOLOGY

C.p. australicus is eurygamous and heterodynamic. Oviposition ceases in April; south of the Dividing Range only the females hibernate, but in the Mildura area larvae occasionally may be found during the winter. The larvae have been found in a variety of habitats both urban and rural. They are particularly numerous in fresh-water swamps and may be found in brackish waters, large and small pools, channels and artificial containers; they are rarely found in completely shaded pools.

C.p. australicus is not a man-biting mosquito; it attacks birds and rabbits.

DISTRIBUTION

Culex pipiens australicus is common in Victoria and occurs in flat country as well as in mountainous country at altitudes up to 5,000 feet. It is widely distributed in Australia and Tasmania; outside Australia it has been recorded only from New Caledonia (Marks and Rageau, 1957).

Culex (Culex) pipiens fatigans Wiedemann

Culex fatigans Wiedemann, 1828, *Ausser. zweifl. Inst.,* 1: 10. *Culex quinquefasciatus* Say, 1823, *Jour. Acad. Nat. Sci. Phila.,* 3: 10. Dobrotworsky and Drummond, 1953, 78: 132.

For additional synonyms see Stone, Knight and Starke (1959).

ADULT FEMALE

Similar to *C.p. australicus* but usually there are no median patches of black scales on the venter; if these are present they are small and inconspicuous. There are no white scales on the postspiracular area.

ADULT MALE

Palps (fig. 84 c) longer than proboscis with labella by terminal segment;

shaft apically less hairy than in *australicus;* terminal segment not directed backwards. *Terminalia* (fig. 84 a, b). Dorsal processes of mesosome pointed and almost parallel; ventral processes broad and leaf-like.

LARVA (fig. 84 f)

Similar to that of *australicus* but has siphon index 3·5-4·8, mean 4·2, and usually only 3 pairs of ventral tufts.

BIOLOGY

C.p. fatigans breeds most abundantly in more or less foul water in

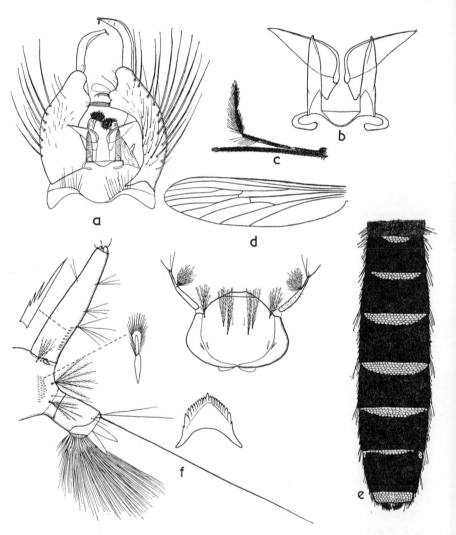

FIG. 84 *Culex pipiens fatigans* Weidemann. *a-e*, adult: *a*, male terminalia; *b*, phallosome; *c*, male proboscis and palp; *d*, female wing; *e*, female abdomen, dorsal view. *f*, larva: head, mentum and terminal segments. (*b*, *c*, after Dobrotworsky.)

domestic situations such as pools, drains and septic tanks, but larvae are also found in water tanks, horse drinking troughs, etc. It is a stenogamous species, but unlike *C.p. molestus*, is anautogenus. Over much of its range in Australia and other parts of the world, *C.p. fatigans* is homodynamous but this does not apply in Victoria. In the northern part of this state, adult activity ceases during the winter, which is usually passed in the larval stage. The species does not seem to be permanently established south of the Central Highlands; in the vicinity of Melbourne, it is quite common in late summer and autumn but neither adults nor larvae have been found during the late winter or spring, suggesting that a reinvasion occurs each year.

C.p. fatigans is a domestic mosquito which readily enters houses and shelters to attack man and domestic poultry at night.

DISTRIBUTION

It occurs throughout the tropics and subtropics of the world and is widely distributed in Australia, but is apparently absent from Tasmania.

Culex (Culex) pipiens molestus Forskal

Culex molestus Forskal, 1775, *Descr. Animalium,* p. 85.
Dobrotworsky and Drummond, 1953, 78: 132.

ADULT FEMALE

It resembles *C. globocoxitus* in having straight tergal bands not constricted laterally but the posterior margins of the bands are not sharply defined and the scales on the tergites are dark brown, not black. The proboscis has dark scales on the underside towards the tip. Two features which serve to distinguish *C.p. molestus* from *C.p. fatigans* and *C.p. australicus* are the straight tergal bands and the short stem of the upper fork cell (fig. 85 d). In *molestus,* the stem is only about one-fifth of the length of the cell, in *fatigans* it is about one-third.

ADULT MALE

The males may be distinguished from those of other Australian members of the *Culex pipiens* complex by their terminalia (fig. 85 a, b). The coxite is not swollen; the paraproct has a basal arm; the ventral processes of the mesosome are narrow and bent outwards; the dorsal processes are stout and bluntly tipped and are directed towards the tips of the ventral processes. It should be noted, however, that in localities where *molestus* and *fatigans* interbreed, a full range of intermediates may be found.

LARVA (fig. 85 f)

This is similar to that of *fatigans,* and there are no reliable morphological traits by which the larvae of *fatigans* and *molestus* can be separated.

EGGS

The egg rafts collected in nature contain from 30 to 180 eggs, those laid autogenously contain from 7 to 135 eggs (mean 57·7).

BIOLOGY

The larvae are found in water butts, in foul water in pools at rubbish tips, in drainage pits, and particularly in septic tanks.

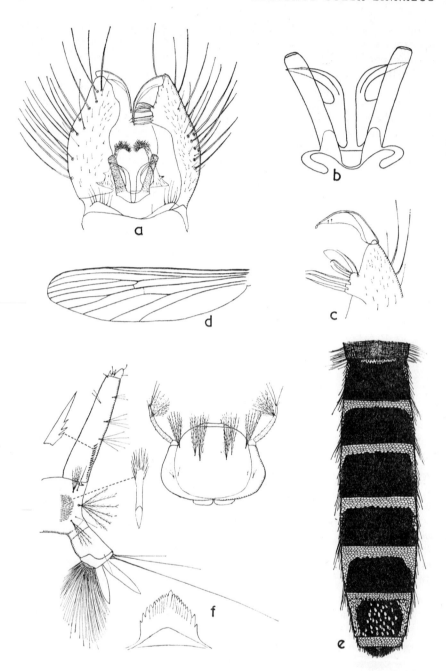

Fig. 85 *Culex pipiens molestus* Forksal. *a-e*, adult: *a*, male terminalia; *b*, phallosome; *c*, part of coxite; *d*, wing; *e*, female abdomen, dorsal view. *f*, larva: head, mentum and terminal segments. (*b*, after Dobrotworsky.)

C.p. molestus is a stenogamous, autogenous and homodynamous mosquito. In Victoria it ceases biting activity during the winter and maintains itself autogenously. In northern Victoria (Mildura area) it continues to feed on blood during the winter and has been recorded as attacking man indoors at temperatures as low as 52°F. It also attacks domestic birds.

DISTRIBUTION

It occurs in the southern parts of the Holarctic Region, in Africa, in the southern part of Australia including Tasmania. In Victoria it is common south of the Dividing Range but to the north is less common than *C.p. fatigans*.

APPENDIX

COLLECTING AND PRESERVING SPECIMENS FOR STUDY

Adults

The most convenient method of collecting adult mosquitoes is by means of a sucking tube (aspirator). This consists of a piece of polystyrene tube, ⅜-½ inch in diameter and about 15 inches long. A disc of fine metal mesh of appropriate diameter is inserted into the tube to act as a screen and then fixed in position by warming the tube over a flame. A rubber tube 2-3 feet long is attached to the end of the polystyrene one (fig. 86 A).

This sucking tube is used for capturing mosquitoes that have settled to feed either on the collector or on birds or mammals that have been exposed as 'bait'. Some species can be collected during the day but the most favoured time is at dusk. Only females will be captured in this way but both sexes can be caught in their resting places, e.g. hollow trees or stumps, vegetation, under ledges or in shelters such as poultry houses.

Another method of collecting, though a less useful one since it often results in damage to the mosquitoes, is by means of a hand net. This is used for sweeping vegetation and for catching mosquitoes that fly out when disturbed in their resting places.

Mosquitoes captured in the sucking tube are blown gently into a chloroform killing tube (fig. 86 B). After 5-10 minutes they may be pinned; if they are left for long periods during which the chloroform tube is being constantly handled, there is a danger of scales being rubbed off.

If the mosquitoes are to be kept alive it is essential, in hot weather, to maintain a high humidity in the cages.

For pinning mosquitoes, steel pins should be used in preference to white ones since the latter tend to develop a green waxy deposit which obscures the specimen. The steel pin is inserted through a piece of polyporus pith mounted on a large pin (No. 16). The mosquito should be laid on its back and the pin inserted between the mid-legs through the thorax. Every care should be taken to avoid brushing off the scales on the body and appendages. Below the specimens the pin should bear a label giving the locality, the date and collector's name.

Adults reared from pupae should not be killed until 15-24 hours after emergence; unless the cuticle has thoroughly hardened the specimens will shrink after pinning.

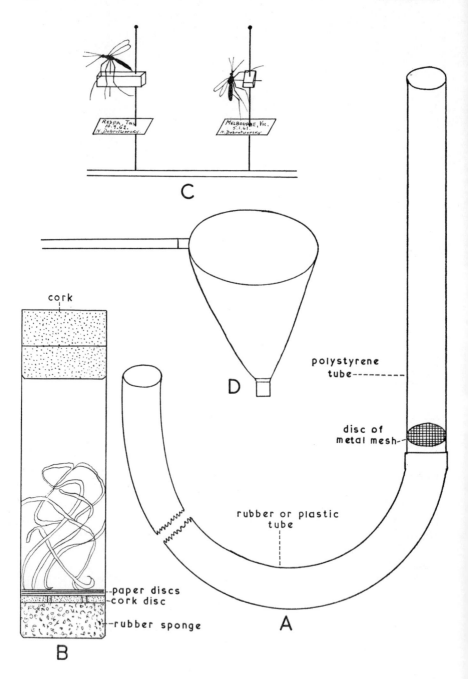

FIG. 86 A, aspirator for collecting adult mosquitoes; B, chloroform tube for killing adult mosquitoes; C, mounted adult mosquitoes; D, net for collecting mosquito larvae.

APPENDIX

Preparation of Male Terminalia

There are several methods of mounting the terminalia but the one recommended by Edwards (1941) is simple and yields permanent mounts. The procedure is: (1) cut off the tip of the abdomen with fine scissors and transfer it to 10 per cent KOH in a test tube, (2) place the tube in a water bath and leave for 5-10 minutes after bringing the bath to boiling point, (3) transfer specimen to water for a few minutes, (4) transfer to glacial acetic acid for 5 minutes, (5) transfer to clove oil for 5 minutes. At this stage the terminalia may be dissected, if necessary. (6) Transfer to a drop of thin Canada Balsam placed near one end of a strip of celluloid about ¼ x ¾ inch, (7) orient the terminalia and set aside to dry, (8) add a small drop of thin balsam and cover with a square of coverslip, (9) impale on pin supporting the specimen to which the terminalia belong.

Larvae

Larvae, and pupae, in pools and swamps can be collected by means of a hand net (fig. 86 d) and then transferred to jars. If the larvae are to be maintained in the laboratory, it is advisable to take a fair quantity of the water in which they are breeding.

Larvae in tree holes are collected by siphoning off the water by a rubber tube; those in leaf axils are collected by means of a pipette. Tree holes containing water are often difficult to find but most tree-hole breeding mosquitoes will oviposit in artificial sites. These are created by placing in hollow trees jars of 1-2 pint capacity filled with water and containing some rotting wood.

Larvae may be preserved in either 70 per cent alcohol or in Macgregor's Solution (borax 5 gm; 40 per cent formalin 100 cc; glycerin 2·5 cc; distilled water to make 1,000 cc). Larvae may be placed directly into the preservative but preferably should first be killed by immersion in hot water. Each sample of larvae should be kept separately in a tube with a label written in lead pencil or indian ink, giving locality, date and name of collector. A record should be kept of the type of breeding place, including information on the state of the water, vegetation, exposure to sun, etc.

REFERENCES

Abbie, A. A., 1941. Marsupials and the evolution of mammals. *Aust. J. Sci.,* 4: 77-92.
Abdel-Malek, A. A. and Baldwin, W. F., 1961. Specificity of plant feeding in mosquitoes as determined by radioactive phosphorus. *Nature,* 192: 178-80.
Anderson, S. C., Dobrotworsky, N. V. and Stevenson, W. J., 1958. Murray Valley Encephalitis in the Murray Valley, 1956 and 1957. *Med. J. Aust.,* July: 15-17.
Barr, A. R., 1960. A review of recent findings on the systematic status of *Culex pipiens*. *Calif. Vect. Views,* 7: 17-21.
Baker, F. C., 1935. The effect of photoperiodism on resting tree-hole mosquito larvae. *Canad. Ent.,* 67: 149-53.
Bates, M., 1949. *The Natural History of Mosquitoes.* New York, The Macmillan Company, 379 pp.
Bekku, H., 1956. Studies on the *Culex pipiens* group of Japan. I. *Nagasaki med. J.,* 31: 956-66.
Belkin, J. N., 1950. A revised nomenclature for the chaetotaxy of the mosquito larva (Diptera: Culicidae). *Amer. Midl. Nat.,* 44: 678-98.
———, 1952. The homology of the chaetotaxy of immature mosquitoes and a revised nomenclature for the chaetotaxy of the pupa (Diptera: Culicidae). *Proc. ent. Soc. Wash.,* 54: 115-30.
———, 1953. Corrected interpretations of some elements of the abdominal chaetotaxy of the mosquito larva and pupa (Diptera: Culicidae). *Proc. ent. Soc. Wash.,* 55: 318-24.
———, 1962. *The Mosquitoes of the South Pacific.* (Diptera: Culicidae). University of California Press, 2 vols, 608 and 412 pp.
Buxton, P. A. and Hopkins, G. H. F., 1927. *Researches in Polynesia and Melanesia.* Parts I-IV. Lond. Sch. Trop. Med., Mem. 1, 260 pp.
Christophers, S. R., 1951. Notes of morphological characters differentiating *Culex pipiens* L. from *Culex molestus* Forskal and the status of these forms. *Trans. R. ent. Soc. Lond.,* 102: 372-9.
Clements, A. N., 1956. Hormonal control of ovary development in mosquitoes. *J. exp. Biol.,* 33: 211-23.
Cooling, L. E., 1924. *A Synomic List of the More Important Species of Culicidae of the Australian Region.* Aust. Dept Health, Serv. Publ. N.2, 61 pp.
Craig, G. B., 1955. Preparation of the chorion of eggs of aedine mosquitoes for microscopy. *Mosquito News,* 15: 228-31.

Davidson, G., 1954. Estimation of the survival-rate of Anophelinae mosquitoes in nature. *Nature*, 174: 792-3.

Day, M. F., Fenner, F. and Woodroofe, G. M., 1956. Further studies on the mechanism of mosquito transmission of myxomatosis in the European rabbit (with an appendix by McIntyre, G. A.). *Hyg. Aust.*, 54: 258-83.

Downes, J. A., 1958. The feeding habits of biting flies and their significance in classification. *Ann. Rev. Ent.*, 3: 249-66.

Dobrotworsky, N. V., 1953. The *Culex pipiens* Group in south-eastern Australia. I. *Proc. Linn. Soc. N.S.W.*, 77: 357-60.

————, 1954a. The Genus *Theobaldia* (Diptera, Culicidae) in Victoria. Ibid., 79: 65-78.

————, 1954b. The *Culex pipiens* Group in south-eastern Australia. III. Autogeny in *Culex pipiens* form *molestus*. Ibid., 79: 193-5.

————, 1955. The *Culex pipiens* Group in south-eastern Australia. IV. Crossbreeding experiments within the *Culex pipiens* Group. Ibid., 80: 33-43.

————, 1956. Notes on Australian mosquitoes (Diptera, Culicidae). I. Some species of the subgenus Neoculex. Ibid., 81: 105-14.

————, 1957. Notes on Australian mosquitoes (Diptera, Culicidae). II. Notes on *Anopheles stigmaticus* Skuse and description of new species of Anopheles from Australia and New Guinea. Ibid., 82: 180-8.

————, 1958. Notes on Australian mosquitoes (Diptera, Culicidae). III. The subgenus *Lophoceraomyia* in Victoria. Ibid., 82: 317-21.

————, 1959. Notes on Australian mosquitoes (Diptera, Culicidae). IV. *Aedes alboannulatus* complex in Victoria. Ibid., 84: 131-45.

————, 1960a. The subgenus *Ochlerotatus* in the Australian Region. III. Review of the Victorian species of Perkinsi and Cunalbulanus Section with description of two new species. Ibid., 85: 53-74.

————, 1960b. The subgenus *Ochlerotatus* in the Australian Region. IV. Review of species of the Flavifrons Section. Ibid., 85: 180-8.

————, 1960c. The genus *Theobaldia* (Diptera, Culicidae) in Victoria. II. Ibid., 85: 240-7.

————, 1960d. Notes on Australian mosquitoes (Diptera, Culicidae). V. Subgenus *Pseudoskusea* in Victoria. Ibid., 85: 257-67.

————, 1962. Notes on Australian mosquitoes (Diptera, Culicidae). VI. Five new Victorian species and a description of the larva of *Aedes milsoni* (Taylor). Ibid., 87: 291-302.

———— and Drummond, F. H., 1953. The *Culex pipiens* Group in south-eastern Australia. II. Ibid., 78: 131-46.

Doherty, R. L., Carley, J. G., Mackerras, M. J., Trevethan, P. and Marks, E. N., 1961. Isolation of Murray Valley Encephalitis and other viruses from mosquitoes in north Queensland. *Aust. J. Sci.*, 23: 302-3.

Doherty, R. L., 1963. Studies of arthropod-borne virus infections in Queensland. III. Isolation and characterization of virus strains from wild-caught mosquitoes in North Queensland. *Austr. J. exp. Biol. med. Sci.*, 41: 17-39.

Douglas, G. W., 1958. Myxomatosis in Victoria. *J. Agric. Vic.*, Dec.: 1-8.

————, 1961. Notes on Australian mosquitoes (Diptera, Culicidae). I. The life history of *Aedomyia venustipes* (Skuse). *Proc. Linn. Soc. N.S.W.*, 86: 262-7.

de Buck, A., Schoute, E. and Swellengrebel, N. H., 1934. Crossbreeding experiments with Dutch and foreign races of *Anopheles maculipennis*. *Riv. Malariol.*, Rome, 13: 237-63.

Drummond, F. H., 1951. The *Culex pipiens* complex in Australia. See Mattingly *et al.* 1951.

Edwards, F. W., 1922. A Revision of the mosquitoes of the Palaearctic Region. *Bull. ent. Res.*, 12: 263-351.

————, 1924. A synopsis of the adult mosquitoes of the Australian Region. *Bull. ent. Res.*, 14: 351-401.

————, 1926. Mosquito Notes. IV. *Bull. ent. Res.*, 17: 101-31.

————, 1932. Diptera. Family Culicidae. *Genera Insect.*, 194: 1-258.

————, 1941. *Mosquitoes of the Ethiopian region*. Lond. British Museum, 499 pp.

Evans, J. W., 1959. The zoogeography of some Australian insects. *Biogeogr. Ecol. Aust., Monogr. Biol.*, 8: 150-63.

Fenner, F., Day, M. F. and Woodroofe, G. M., 1952. The mechanism of the transmission of myxomatosis in the European rabbit (*Oryctolagus cuniculus*) by the mosquito *Aedes aegypti*. *Aust. J. exp. Biol. med. Sci.*, 30: 139-52.

Gjullin, C. M., Hegarty, C. P. and Bollen, W. B., 1941. The necessity of a low oxygen concentration for the hatching of *Aedes* mosquito eggs. *J. Cell. Comp. Physiol.*, 17: 193-202.

Harrison, L., 1928. The composition and origins of the Australian fauna, with special reference to the Wegener Hypothesis. 18th Meeting Aust. Ass. Adv. Sci., Report, 332.

Hill, G. F., 1925. The distribution of Anopheline mosquitoes in the Australian Region, with notes on some Culicine species. *Proc. Roy. Soc. Vic.*, 37: 61-7.

Horsfall, W. R., Miles, R. C. and Skoatch, J. T., 1952. Eggs of floodwater mosquitoes. I. Species of *Psoropora*. *Ann. ent. Soc. Amer.*, 45: 618-24.

Kamura, T., 1958. Studies on the *Culex pipiens* group of Japan. III. On the seasonal changes of morphological characters in Isahaya *pallens*. *Nagasaki med. J.*, 33: 78-85.

Key, K. H. L., 1959. The ecology and biogeography of Australian grasshoppers and locusts. *Biogeogr. Ecol. Aust., Monogr. Biol.*, 8: 192-210.

Kitzmiller, J. B., 1953. Mosquito genetics and cytogenetics. *Rev. brazil Malariol*, 5: 285-359.

————, 1959. Race formation and speciation in mosquitoes. *Cold Spring Harb. Symp. quant. Biol.*, 24: 161-5.

———— and Laven, H., 1959. Current concepts of evolutionary mechanism in mosquitoes. *Cold Spring Harb. Symp. quant. Biol.*, 24: 173-5.

Lane, J., 1939. *Catalogue des Mosquitoes Neotropicos*. Bol. biol. Cl. zool. Brasil (Monogr.), 218 pp.

Laven, H., 1951a. See Mattingly *et al.*, 1951.
———, 1951b. Crossing experiments with *Culex* strains. *Evolution*, 5: 370-5.
———, 1953. Reziprok unterschiedliche Kreuzbarkeit von Steckmücken (Culicidae) und ihre Deutung als plasmatishe Vererbung. *Z. ind. Abs. Vererb.*, 85: 118-36.
———, 1957. Vererbung durch Kerngene und das Problem der ansserkaryotischen Vererbung bei *Culex pipiens*. I. Kernvererbung. *Zeit induk. Abstam. u. Vererb-lehre*, 88: 443-77.
———, 1959. Speciation by cytoplasmic isolation in the *Culex pipiens* complex. *Cold Spring Harb. Symp. quant. Biol.*, 24: 166-73.
Lee, D. J., 1944. *An Atlas of the Mosquito Larvae of the Australian Region. Tribes Megarhini and Culicini*. Melbourne. Australian Military Forces.
———, Clinton, K. J. and O'Gower, A. K., 1954. The blood sources of some Australian mosquitoes. *Austr. J. biol. Sci.*, 7: 283-301.
———, Dyce, A. L. and O'Gower, A. K., 1957. Blood-sucking flies (Diptera) and myxomatosis transmission in a mountain environment in New South Wales. *Austr. J. Zool.*, 5: 355-401.
———, Fenner, F. and Lawrence, J. J., 1958. Mosquitoes and fowl pox in Sydney area. *Austr. vet. J.*, August: 230-7.
———, Woodhill, A. R., 1944. *The Anopheline Mosquitoes of the Australian Region*. Zool. Dept. Univ. Sydney, Monogr. 2, 209 pp.
Mackerras, I. M., 1927. Notes on Australian mosquitoes (Diptera, Culicidae). Part II. The zoogeography of the subgenus Ochlerotatus with notes on the species. *Proc. Linn. Soc. N.S.W.*, 52: 284-98.
———, 1950. The Zoogeography of the Diptera. *Aust. J. Sci.*, 12: 157-61.
McLean, D. M., 1953. Transmission of Murray Valley Encephalitis virus by mosquitoes. *Aust. J. exp. Biol. med. Sci.*, 31: 481-90.
Marks, E. N., 1949. Studies of Queensland mosquitoes. IV. Some species of *Aedes* (subgenus *Ochlerotatus*). *Pap. Dep. Ent. Univ. Qd*, 2 (11): 1-41.
———, 1954. Research on Australian mosquitoes. Presidential Address, Entomological Society of Queensland.
———, 1955. Studies of Queensland mosquitoes. V. Some species of *Aedes* (subgenus *Finlaya*). *Pap. Dep. Ent. Univ. Qd*, 1 (2): 11-29.
———, 1957. The Subgenus *Ochlerotatus* in the Australian Region (Diptera: Culicidae). I. Notes on classification, with the description of a new species. *Pap. Dept. Ent. Univ. Qd*, 1 (5): 71-83.
———, 1958. New species and records of the *Aedes kochi* Group from eastern Australia (Diptera: Culicidae). *Proc. Roy. Soc. Qd*, 19: 57-74.
———, 1960. Mosquitoes biting frogs. *Aust. J. Sci.*, 23: 89.
———, 1961. Faunal relationships of some Australian and Papuan Culicidae. XI. *Intern. Kongr. Entom. Wien. 1960*, 1: 185-7.
———, 1963a. A revision of the subgenus *Chaetocruiomyia* Theobald (Diptera: Culicidae). *Pap. Dep. Ent. Univ. Qd*, 1 (13): 189-211.
———, 1963b. The subgenus *Ochlerotatus* in the Australian Region

REFERENCES

(Diptera: Culicidae). V. The Stricklandi Section. *J. ent. Soc. Qd*, 2: 31-47.

——————, 1964. Notes on the subgenus *Chaetocruiomyia* Theobald (Diptera: Culicidae). *Proc. Linn. Soc. N.S.W.*, 89: 131-47.

——————, and Rageau, J., 1957. *Culex pipiens australicus* Dobrotworsky and Drummond in New Caledonia. *Proc. Linn. Soc. N.S.W.*, 82: 156.

Matheson, R., 1944. *A Handbook of the Mosquitoes of North America.* Ithaca, N.Y., 314 pp.

Mattingly, P. F., Rozenboom, L. E., Knight, K. L., Laven, H., Drummond, F. H., Christophers, S. R. and Shute, P. G., 1951. The *Culex pipiens* complex. *Trans. R. ent. Soc. Lond.*, 102: 331-82.

Mayr, E., 1942. *Systematics and the Origin of Species.* Colombia University Press, 334 pp.

Muspart, J., 1953. Research on South African Culicini (Diptera, Culicidae). I. Descriptions of two new *Aedes* (*Ochlerotatus*) Lynch Arribalzaga. *J. ent. Soc. S. Afr.*, 16: 51-8.

Mykytowycz, R., 1956. The effect of season and mode of transmission on the severity of myxomatosis due to an attenuated strain of the virus. *Aust. J. exp. Biol.*, 34: 121-32.

Nye, E. R., 1962. *Aedes* (*Pseudoskusea*) *australis* Erichson (Diptera: Culicidae) in New Zealand. *Trans. Roy. Soc. N.Z.*, Zool. 3: 33.

O'Gower, A. K., 1958. The oviposition behaviour of *Aedes australis* (Erichson) (Diptera, Culicidae). *Proc. Linn. Soc. N.S.W.*, 83: 245-50.

Omory, N., Bekker, H. and Kamura, T., 1955. On the occurrence of *Culex pipiens molestus* in Nagasaki City (preliminary report). *Nagasaki med. J.*, 30: 1572-6.

Paramonov, S. J., 1959. Zoogeographical aspects of the Australian Dipterofauna. *Biogeogr. Ecol. Aust., Monogr. Biol.*, 8: 164-91.

Patton, R. T., 1955. Vegetation in Victoria. *Introducing Victoria,* ed. Leeper, G. W., Melbourne University Press, 52-68.

Pillai, J. S., 1962a. A celloidin impression technique for recording egg sculpturing in *Aedes* mosquitoes. *Nature,* 194: 212-13.

——————, 1962b. Factors influencing egg survival in *Aedes* eggs with special reference to some Victorian species (Diptera, Culicidae). Ph.D. thesis, Zool. Dept. Univ. Melbourne.

Ratcliffe, F. N., Myers, K., Fennessy, B. V. and Calaby, J. H., 1952. Myxomatosis in Australia. *Nature,* 170: 7-11.

Reeves, W. C., French, E. L., Marks, E. N. and Kent, N. E., 1954. Murray Valley Encephalitis: a survey of suspected mosquito vectors. *Am. J. trop. Med. Hyg.*, 3: 147-59.

Roberts, F. H. S., 1943. Observations on *Anopheles annulipes* Walk. as a possible vector of malaria. *Aust. J. exp. Biol. med. Sci.*, 21: 259-62.

Roubaud, E. and Colas-Belcour, J., 1927. Action des diastases dans le determinisme d'éclosion de l'oeuf chez le moustique de la fièvre jaune (*Stegomyia fasciata*). *Acad. des. Sci., Compt. Rend.*, 184: 248-9.

Rozeboom, L. E. and Gilford, B. N., 1954. Sexual isolation between populations of the *Culex pipiens* complex in North America. *J. Parasit.*, 40: 237-44.

———— and Kitzmiller, J. B., 1958. Hybridization and speciation in mosquitoes. *Ann. Rev. Ent.*, 3: 231-48.

Sandholm, H. A. and Price, R. D., 1962. Field observations on the nectar feeding habits of some Minnesota mosquitoes. *Mosquito News*, 22: 346-51.

Serventy, D. L. and Whittell, H. M., 1948. *A Handbook of the Birds of Western Australia*. Perth, Patersons Press, 365 pp.

Stone, A., Knight, K. L. and Starcke, H., 1959. *A Synoptic Catalog of the Mosquitoes of the World (Diptera, Culicidae)*, Ent. Soc. Wash. (Thomas Say Foundation), 358 pp.

Stone, A., 1961. Supplement I to *A Synoptic Catalog of the Mosquitoes of the World. Proc. ent. Soc. Wash.*, 63: 29-52.

Strickland, E. H., 1911. Some new *Culicidae* from Western Australia, South Queensland and Tasmania. *Entom.*, 44: 130-4, 179-182, 201-4.

Swellengrebel, N. H., de Buck, A. and Schoute, E., 1927. On Anophelism without malaria round Amsterdam. *Proc. Kon. Akad. Wetensch.*, Amsterdam, 30: 61-8.

Taylor, F. H., 1917. *Malaria Mosquito Survey of Irrigation Areas in the Murray River District*. Aust. Quarant. Service, Serv. Publ. 12, Melbourne, 32 pp.

Twohy, D. W. and Rozeboom, L. E., 1957. A comparison of the food reserves in autogenous and anautogenous *Culex pipiens* populations. *Am. J. Hyg.*, 65: 316-24.

Wallis, R. C., 1955. A study of the oviposition activity of three species of *Anopheles* in the laboratory. *Am. J. trop. Med. Hyg.*, 4: 557-63.

West, A. S. and Jenkins, D. W., 1951. Plant feeding habits of northern mosquitoes studied with radio-isotopes. *Mosquito News*, 11: 217-19.

Woodhill, A. R., 1936. Observations and experiments on *Ae. concolor* Taylor. *Bull. ent. Res.*, 27: 633-48.

SYSTEMATIC LIST OF MOSQUITO SPECIES

Family CULICIDAE

Subfamily ANOPHELINAE	40
Genus *Anopheles* Meigen	41
Subgenus *Anopheles* Meigen	41
stigmaticus Skuse	41
pseudostigmaticus Dobrotworsky	42
atratipes Skuse	45
Subgenus *Cellia* Theobald	47
annulipes Walker	47
Subfamily CULICINAE	50
Tribe Sabethini	51
Genus *Tripteroides* Giles	51
Subgenus *Rachionotomyia* Theobald	52
Group *atripes*	52
atripes Skuse	52
southern form of *atripes* Skuse	53
Group *caledonicus*	55
tasmaniensis Strickland	56
Group *argenteiventris*	58
marksae Dobrotworsky	58
Tribe Culicini	60
Genus *Mansonia* Blanchard	60
Subgenus *Coquillettidia* Dyar	61
linealis Skuse	61
aurata Dobrotworsky	63
variegata Dobrotworsky	63
Subgenus *Mansonioides* Theobald	65
uniformis Theobald	66
Genus *Aedeomyia* Theobald	67
venustipes Skuse	68
Genus *Aedes* Meigen	70
Subgenus *Mucidus* Theobald	74
alternans Westwood	74

Subgenus *Ochlerotatus* Lynch Arribalzaga	76
Vigilax Section	79
vigilax Skuse	79
procax Skuse	81
Theobaldi Section	83
theobaldi Taylor	83
theobaldi eidsvoldensis Mackerras	85
Burpengaryensis Section	86
nigrithorax Macquart	86
sagax Skuse	88
vittiger Skuse	91
imperfectus Dobrotworsky	93
Flavifrons Section	95
flavifrons Skuse	95
calcariae Marks	97
purpuriventris Edwards	100
clelandi Taylor	102
Perkinsi Section	104
perkinsi Marks	104
luteifemur Edwards	106
silvestris Dobrotworsky	109
nivalis Edwards	112
Bogong form of *nivalis* Edwards	114
camptorhynchus Thomson	115
Cunabulanus Section	117
andersoni Edwards	117
Grampians form of *andersoni* Edwards	118
continentalis Dobrotworsky	120
Stricklandi Section	122
stricklandi Edwards	123
spilotus Marks	125
Subgenus *Finlaya* Theobald	127
Group *kochi*	128
dobrotworskyi Marks	128
Group *mediovittatus*	131
notoscriptus Skuse	131
mallochi Taylor	133
plagosus Marks	135
Group *alboannulatus*	136
alboannulatus Macquart	137
rubrithorax Macquart	140
rupestris Dobrotworsky	143

tubbutiensis Dobrotworsky	146
subbasalis Dobrotworsky	148
milsoni Taylor	150
Group undetermined	152
subauridorsum Marks	152
Subgenus *Macleaya* Theobald	154
tremulus Theobald	154
Subgenus *Chaetocruiomyia* Theobald	157
wattensis Taylor	157
macmillani Marks	159
Subgenus *Pseudoskusea* Theobald	161
bancroftianus Edwards	161
postspiraculosis Dobrotworsky	163
multiplex Theobald	166
Subgenus *Halaedes* Belkin	168
australis Erichson	168
Genus *Culiseta* Felt	170
Subgenus *Austrotheobaldia* Dobrotworsky	172
littleri Taylor	172
Subgenus *Culicella* Felt	174
victoriensis Dobrotworsky	175
drummondi Dobrotworsky	177
sylvanensis Dobrotworsky	179
otwayensis Dobrotworsky	181
inconspicua Lee	183
Subgenus *Climacura* Howard, Dyar and Knab	184
antipodea Dobrotworsky	185
Subgenus *Neotheobaldia* Dobrotworsky	187
hilli Edwards	187
frenchii Theobald	189
frenchii atritarsalis Dobrotworsky	191
Genus *Culex* Linnaeus	191
Subgenus *Neoculex* Dyar	193
Group *apicalis*	193
fergusoni Taylor	194
douglasi Dobrotworsky	196
postspiraculosus Lee	198
pseudomelanoconia Theobald	200
Subgenus *Lophoceraomyia* Theobald	202
Group *minutissimus*	202
cylindricus Theobald	202

Group *fraudatrix* 204
   orbostiensis Dobrotworsky 204
Subgenus *Culex* Linnaeus 206
  Group *sitiens* 207
   annulirostris Skuse 207
  Group *pipiens* 209
   globocoxitus Dobrotworsky 211
   pipiens australicus Dobrotworsky and Drummond 213
   pipiens fatigans Wiedemann 215
   pipiens molestus Forskal 217

INDEX

For suprageneric names see the Systematic List of Mosquito Species. Valid genera and subgenera are in **bold type**; species and subspecies in non-bold; all synonyms in *italics*. Bold numerals indicate main references.

Aedeomyia Theob. 5, 50, 51, **67-70**
Aedes Meigen 10, 13, 15, 16, 17, 18, 20, 26, 33, 34, 50, 51, **70-170**
Aedimorphus Theob. 154
Aedomyia Giles 67
albescens Taylor 15
aegypti (Linn.) 4
albirostris Macq. 79
alboannulata Taylor 154
alboannulatus (Macq.) 34, 35, 73, 136, **137-40**, 143, 146
alternans Westw. 3, 26, 31, 71, **74-6**
andersoni Edw. 15, 31, 73, 78, **117-20**
Andersonia Strickl. 117
annulipes Walker 3, 31, 33, 35, 36, 40, 41, **47-9**
annulipes Taylor 115
annulirostris Skuse 14, 28, 31, 35, 36, 192, 193, **207-9**
Anopheles Meigen 3, 5, 13, 40, **41-9**
antipodea Dobrot. 171, 172, **185-6**
apicalis Adams 32, 33
apicotriangulata Theob. 52
atratipes Skuse 36, 40, 41, **45-7**
atripes (Skuse) 13, 31, **52-5**
atritarsalis Dobrot. 191
aurata Dobrot. 15, 31, **63**
australicus Dobrot. & Drumm. 14, 15, 16, 33, 34, 35, 193, 210, **213-15**
australiensis Giles 66
australis (Erich.) 3, 14, 15, 17, 39, 72, **168-70**
australis Strickl. 123
australis Taylor 154
australis Theob. 112
Austrotheobaldia Dobrot. 171, **172-4**

bancroftianus Edw. 26, 31, 33, 71, **161-3**

Caenocephalus Taylor 168
calcariae Marks 73, 77, **97-9**, 104
camptorhynchus (Thom.) 31, 33, 35, 73, 77, **115-17**
Cellia Theob. 40, **47-9**
cephasi Edw. 56

Chagasia Cruz. 40
Chaetocruiomyia Theob. 71, **157-60**
Chrysoconops Goeldi 172
clelandi (Taylor) 73, 78, 102
Climacura Howard, Dyar and Knab 171, **184-6**
Colonemyia Leic. 51
commovens Walker 74
concolor Taylor 168
Coquillettidia Dyar 60, **61-5**
consimilis Taylor 207
continentalis Dobrot. 73, 78, **120-2**
crucians Walker 168
Culex Linn. 5, 13, 18, 26, 32, 50, 51, **191-219**
Culicada Felt 95, 102, 115, 150
Culicelsa Felt 115, 140, 207
Culiseta Felt 5, 7, 11, 13, 16, 18, 26, 27, 31, 32, 33, 34, 35, 50, 51, **170-91**
Culicella Felt **174-84**
cumpstoni Taylor 140
cunabulanus Edw. 74, 78, 120
Cyathomyia Meijere 202
cylindricus Theob. 193, **202-4**

Danielsia Theob. 154
darwini Taylor 154
demansis Strickl. 140
derricki Taylor 47
dobrotworskyi Marks 16, 27, 71, **128-31**
doddi Taylor 154
douglasi Dobrot. 34, 192, **196-8**
drummondi (Dobrot.) 172, **177-9**, 181

eidsvoldensis Macker. **85-6**
Ekrinomyia Leic. 74
Eumelanomyia Theob. 193

farauti Laver. 36
fatigans Wied. 14, 33, 193, 209, 210, **215-17**, 219
fergusoni (Taylor) 15, 31, 32, 34, 192, 193, **194-6**
flindersi Taylor 123
Finlaya Theob. 18, 34, 71, **127-54**
flavifrons (Skuse) 17, 73, 77, **95-7**, 102
frenchii (Theob.) 171, 172, **189-91**

gambiae Giles 15
globocoxitus Dobrot. 14, 15, 33, 192, 193, 210, **211-13**
Grabhamia Theob. 83, 123

Halaedes Belkin 71, **168-70**
hilli (Edw.) 14, 15, 171, 172, **187-9**
hispidosus Skuse 74
Hulecoeteomyia Theob. 150
hybrida Taylor 140

imperfectus Dobrot. 20, 72, 78, **93-5**
inconspicua (Lee) 15, 171, 172, **183-4**
inornata Strickl. 115

jepsoni Bahr 207

labeculosus Coquil. 115
Lepiothauma Enderl. 67
linealis (Skuse) 15, **61-3**
littleri (Taylor) 171, **172-4**
Lophoceraomyia Theob. 15, **202-6**
Lophoceratomyia Theob. 202
luteifemur Edw. 31, 74, 78, **106-9**

macleayanus Macker. 86
Macleaya Theob. 71, **154-6**
macmillani Marks **159-61**
Maillotia Theob. 193
mallochi Taylor 72, **133-5**
Mansonia Blanch. 13, 18, 23, 26, 28, 50, 51, **60-7**
Mansonioides Theob. 60, 61, **65-7**
marksae Dobrot. **58-60**
mastersi Skuse 47
Menolepsis Lutz. 112, 117
milsoni (Taylor) 73, 136, 137, **150-2**
Mimeteomyia Theob. 52, 154
minuta Taylor 154
molestus Forsk. 14, 15, 27, 33, 192, 193, 209, **217-19**
Mucidus Theob. 5, 71, **74-6**
multiplex (Theob.) 31, 72, **166-8**
musivus Skuse 47

Neoculex (Dyar) **193-202**
Neotheobaldia Dobrot. 171, **187-91**
nigra Taylor 115
nigrithorax (Macq.) 35, 72, 78, **86-8**, 89
nivalis Edw. 31, 50, 74, 77, 78, **112-15**
notoscriptus (Skuse) 20, 27, 72, **131-3**

Ochlerotatus Lyn. Arrib. 13, 20, 31, 32, 34, 71, **76-127**
orbostiensis Dobrot. 13, 16, 31, 193, **204-6**
otwayensis (Dobrot.) 171, 172, **181-2**

palpalis Taylor 207
Panoplites Theob. 60

Pardomyia Theob. 74
perkinsi Marks 32, **104-6**
perplexus Taylor 47
persimilis Taylor 47
Philodendromyia Theob. 202
pipiens Linn. 15, 35, 209, 210
plagosus Marks **135-6**
Polylepidomyia Theob. 52
postspiraculosis Dobrot. 14, 15, 26, 31, 72, **163-6**
postspiraculosus Lee 192, 193, **198-200**
procax Skuse 31, 73, 77, **81-3**
Protomelanoconion Theob. 193
pseudomelanoconia Theob. 31, 192, 193, **200-2**
Pseudoskusea Theob. 34, 71, **161-8**
pseudostigmaticus Dobrot. 15, 40, 41, **42-5**
Pseudotaeniorhynchus Theob. 60
Pseudotheobaldia Theob. 170
pulcherrimus Taylor 133
purpuriventris Edw. 31, 73, 78, **100-2**

queenslandis (Strickl.) 136, 140
quinquefasciatus Say 215

Rachionotomyia Theob. **52-60**
Rhynchotaenia Brèthes 60
rubrithorax (Macq.) 16, 20, 31, 73, 136, 137, **140-3**, 146
rupestris Dobrot. 16, 20, 27, 31, 72, 136, 137, **143-6**, 145, 148

sagax (Skuse) 20, 31, 32, 33, 34, 72, 78, **88-90**
silvestris Dobrot. 31, 74, 77, 78, **109-12**
similis Strickl. 140
simplex Taylor 207
Skeiromyia Leic. 52
Skusea Theob. 166
somerseti Taylor 207
spilotus Marks 72, 78, **125-7**
Squamomyia Theob. 52
Stegomyia Theob. 56, 131
stigmaticus Skuse 15, 31, 40, **41-2**
stricklandi (Edw.) 72, 78, **123-5**
subauridorsum Marks 73, 128, **152-4**
subbasalis Dobrot. 31, 73, 136, 137, **148-50**
sylvanensis (Dobrot.) 171, 172, **179-181**

Taeniorhynchus Lyn. Arrib. 60, 61
tasmaniensis (Strickl.) 14, 15, 26, 31, **56-9**, 117, 168
tasmaniensis Taylor 112, 117
theobaldi (Taylor) 20, 26, 31, 33, 34, 72, 78, **83-6**, 163
Theobaldia Neveu-Lem. 170
Theobaldinella Blanch. 170

INDEX

tremulus (Theob.) 72, **154-6**
Tricholeptomyia Dyar and Shannon 52
Tripteroides Giles 5, 7, 13, 16, 18, 27, 34, 50, **51-60**
tubbutiensis Dobrot. 73, 136, 137, **146-8**

Uranotaenia Lyn. Arrib. 15
uniformis (Theob.) 66-7

vandema Strickl. 95
variegata Dobrot. 31, **63-5**

venustipes (Skuse) 15, **68-70**
victoriensis Taylor 115, 172
victoriensis (Dobrot.) 172, **175-7**
vigilax (Skuse) 28, 31, 36, 72, 78, **79-81**
vittiger (Skuse) 20, 28, 31, 32, 33, 34, 72, 77, **91-3**

waterhousei Dobrot. 109
wattensis Taylor **157-9**
westralis Strickl. 115
wilsoni Taylor 88